PoliTO Springer Series

Springer, in cooperation with Politecnico di Torino, publishes the PoliTO Springer Series. This co-branded series of publications includes works by authors and volume editors mainly affiliated with Politecnico di Torino and covers academic and professional topics in the following areas: Mathematics and Statistics, Chemistry and Physical Sciences, Computer Science, All fields of Engineering. Interdisciplinary contributions combining the above areas are also welcome. The series will consist of lecture notes, research monographs, and briefs. Lectures notes are meant to provide quick information on research advances and may be based e.g. on summer schools or intensive courses on topics of current research, while SpringerBriefs are intended as concise summaries of cutting-edge research and its practical applications. The PoliTO Springer Series will promote international authorship, and addresses a global readership of scholars, students, researchers, professionals and policymakers.

More information about this series at http://www.springer.com/series/13890

Giacomo Chiesa

Technological Paradigms and Digital Eras

Data-driven Visions for Building Design

POLITECNICO
DI TORINO

Springer

Giacomo Chiesa
Dipartimento di Architettura e Design (DAD)
Politecnico di Torino
Turin, Italy

ISSN 2509-6796 ISSN 2509-7024 (electronic)
PoliTO Springer Series
ISBN 978-3-030-26198-6 ISBN 978-3-030-26199-3 (eBook)
https://doi.org/10.1007/978-3-030-26199-3

This Springer imprint is published by the registered company Springer Nature Switzerland AG
The registered company address is: Gewerbestrasse 11, 6330 Cham, Switzerland

This book is dedicated to Tat and Chloe.

Foreword

A current trend in the construction sector is designing and operating buildings increasingly "sustainable" and "smart", responding to worldwide, European[1] and national standards based on the assessment and certification of environmental, social, and economic performance. This trend is as well responding to market needs for invigorating an otherwise almost dying demand for new constructions. This situation poses new challenges to architects, engineers and builders as well as any other operator of the construction sector.

In particular, these challenges can be characterised mainly by the following aspects.

- An accurate and detailed knowledge of the physical context of the building under design, including climate, geomorphology, energy network and material flows.
- Availability of calculation and simulation tools able to model and predict the building's performance in terms of environmental impacts and functional operation efficiency during its expected lifespan. Such tools shall have, differently from the majority of the mostly diffused current ones, a parametrical and holistic approach to allow designers for checking while designing, since the earliest stages, the effect of any choice on the future performance and, reciprocally, evaluating formal options as a consequence of specific environmental and functional requirements.
- Availability of hardware, and relevant operating knowledge, concerning the smart control of a building, automatically and by its users, in order to make its use the most efficient in terms of energy consumption and environmental impact.

[1]See, in particular, the series of standards developed within the CEN TC 350 working groups on the assessment of sustainability in buildings and engineering constructions at both levels of building and product.

The present publication by Giacomo Chiesa offers answers to all these aspects, through analyses, methods and tools, which are original and based on a scientifically robust background.

His contribution includes an investigation of the state of the art of theories and methodologies, on which elaboration and management of data for urban and building design are based. Within this framework, he sets guidelines for a future perspective, within which building performance simulation during the design process, with particular reference to energy and environmental aspects, is characterised by a new approach, non-deterministic, but circular and based on probabilistic models. In addition, a thorough analysis of procedures and instruments used to control the environmental indoor conditions of a building during operation is carried out, considering the most advanced application of AI and IoT.

Turin, Italy

Mario Grosso
Full Professor (Retired)
Architectural Technology and
Environmental Design

Preface

ICT and IT are primary sources of technological innovation in the building sector and in design practice which suggests the need to change methodological approaches and tools used in order to move towards feedback-based and explicit visions. It should be underlined that only with the new generation of instruments has it been possible to fully apply informatics-based approaches to design, even if this possibility has already been part of innovative visions since the 50s of the XX Century—see, for example, the design vision at the base of the Ulm School of Desing (1953–1968) and consider Maldonado's observations.

The proposed book is principally based on the performance-driven approach according to the Italian methodological approach (see Chap. 4). This method, named "metodo esigenziale-prestazionale", was discussed and partially regulated thanks to the work of researches principally from the architectural technology research field[2]. Nevertheless, even if the concepts of algorithmic design and programme-driven design were theorised in that period, only today can the applications of this method be brought fully into force thanks to higher computational power, access to new data and the presence of new tools including the connection between geometries and data (e.g. BIM, DIM and CIM). Furthermore, these new instruments require a revision and transformation of early prodromic visions into reliable approaches which are in line with ICT potentials and which are able to include future trends and possibilities.

This book aims at giving to architects a framework to read, interpret and be ready for the digital revolution in terms of re-design and re-engineering of current methods, tools and strategies. Furthermore, it aims at connecting the ICT world and its language with the world of architecture, and for this reason, it is also addressed to IT engineers and other experts who aim to enter the world of building. This is a new frontier in research, education and at different professional levels as has been

[2] See, for example, Cavaglià G, Ceragioli G, Foto M, Maggi PN, Matteoli L, Ossola F (1975) Industrializzazione per programmi. Strumenti e procedure per la definizione dei sistemi di edilizia abitativa. Studi e Ricerche RDB, Piacenza; and Ciribini G (1968) *Brevi noti di metodologia della progettazione architettonica*. Edizioni quaderni di studio, Politecnico di Torino, Torino.

underlined during innovative courses at university level, in several EU-funded research schemes, and in the increasing number of professional publications about smart/digital building and urban applications.

For these reasons, the book is principally devoted to professionals and early stage researchers even if it can also be useful for master degree students, from both building and ICT domains, as it aims at providing, on the one hand, a theoretical background and, on the other, simple sample applications of the hybridisation of these two domains. A method to interpret the current state of the art is, in fact, provided together with practical examples which are easy to understand and can be replicated by interested readers who would like to enter the digital paradigm in connection with the building world. The book is original because examples are not fully applied and interpreted to a specific case. Rather, they are conceived to relate visions to specific examples which are focused on the presentation of the potentialities of technological innovations while considering different aspects of hybridisation (eras), from the modelling to the materialisation worlds, including data-driven innovations (such as big data analyses) while giving readers the scope to adapt potentialities to all proprietary or open platforms without focusing only on a limited number of them.

Parts of the contents of this book were the subject of a previous publication—Paradigmi, tecnologie ed ere digitali—published by Accademia University Press, Turin, in 2015—while others were derived from my Ph.D. final dissertation—Monitoring Energy and Technological Real-time data for Optimisation (M.E.T.R.O.) innovative responsive conception for cityfutures—discussed in 2014 (Matteoli and Pagani 2010). Nevertheless, the manuscript has been totally upgraded and new chapters included.

The book is introduced in Chap. 1, followed by a state of the art in Chap. 2, which outlines the proposed methodological framework and organises references according to the highlighted digital innovation spheres (eras). Chapter 3 discusses the era of modelling, connecting real and virtual environments by discussing, a simple IoT application connected with a CAD environment, while Chap. 4 introduces the relation between the proposed vision and the performance-driven approach by analysing potential applications in shape parametric generation and optimisation according to environmental criteria. Furthermore, the idea of materialisation is also discussed thanks to the possibility to panel a surface to be produced by using computer numerical controlled machines. Chapter 5 is devoted to a brief presentation of the implications of diffuse sensing and big data analysis by focusing on a simple example of environmental diffuse monitoring nodes. Finally, Chap. 6 returns to the theory presented earlier including the conceptualisation of the platform idea, while Chapter 7 presents the most recently described digital implication (the Third Era) connected to data. In this chapter discussed data social implications are also briefly discussed.

Enrico Casetta is thankfully acknowledged for co-authoring Chap. 4.

Turin, Italy Giacomo Chiesa

Introduction

Background

Over the last 60 years, the world has changed radically. Following innovations in Information and Communication Technologies (ICT), the potential of design is changing thanks to new ways of reading in the field of computer hardware and software. Production techniques are rapidly opening up to a widespread production perspective through mass customisation techniques, sometimes of an amateur nature, such as open-source development environments characterised by bottom-up forces, implemented by hobbyists and makers. This open-source attention highlights the overcoming, in certain areas, of that stated in Gates' letter dated 3 February 1976, in which the founder of Microsoft accused hobbyists of stealing his software ("Most directly, the thing you do is theft"), effectively preventing the development of quality programs. In fact, despite the statements "Hardware must be paid for, but software is something to share. Who cares if the people who worked on it get paid?", the Linux operating system enjoyed extensive development and a large number of implementations, by experts and hobbyists around the world, due precisely to its free and unrestricted distribution. Similarly, many other programming tools and languages can be used free of charge, such as R, a "free software environment for statistical computing and graphics"—see also https://www.r-project.org/, viewed December 2018—and the more and more diffused Python, a programming language characterised to be quick and integrated—see also https://www.python.org/, viewed December 2018. At the same time, we have now entered the Third Industrial Revolution, as Anderson calls it, based on the concept of "give away the bits and charge for the atoms". Moreover, way back in 1950, Wiener, the father of cybernetics, warned against secreting data and information because the disclosure of every secret is only a matter of time and every discovery in this sense increases the need for a new and more powerful discovery.

Reflecting on the ownership of software and hardware would, however, be pointless if it did not have immediate repercussions on data revolution. In fact, it is possible to highlight the links with big data, open-source systems for rapid prototyping and the new widespread interest in the production of components and software in the world of the Internet of things and smartness. Although the real revolution does not lie in the machines, but in the data, as we are reminded by Mayer-Schönberger and Cukier, IoT platforms can lead to predominant control by their owner (see also Pfister 2011). Possessing data and technologies risks increasing "the human use of human beings", from a point of view where people are used as the workings of a machine, losing the use of the full faculties of human beings and risking not being able to work out the right answers to our questions, as introduced by Wiener. At the same time, there is a scientifically proven need to further increase knowledge and skills in order to manage and popularise complex decision-making models, capable of assessing and optimising certain choices aimed at increasingly systemic and eco-systemic sustainability.

Topics

The city is becoming a complex reality and a place of hybridisation between the real and virtual worlds. Information and data, relating to complex urban and architectural processes in bidirectional flows, have become parameters of design and civilisation.

In this context, predictive models, quality analyses and process management are undergoing major innovation due to the dissemination of information and data flows from sensors positioned in the real world such as to allow real-time responses. The traditional scarcity of data gives way to an abundance of information that can be generated by sensors applicable to every person and every object. Managing the complexity of networks and smart cities interfaces with techniques for the creation, processing and use of data. At the same time, traditional ways of producing data through monitoring campaigns are being enriched by new potentials deriving from the use of widespread sensors and datapoints, which are less precise in terms of single measurement but allow an increase in the frequency and size of data sets. This implication becomes particularly important in the age of the Internet 3.0 (the Internet of things). Innovations in ICT concern the intrinsic potential of using large data sets to improve the performance of simulation, forecasting, modelling and materialisation systems from a perspective linked increasingly to hybrid (real–virtual) worlds and operating rating assessments. In this sense, these systems are configured as tools and information to respond to the increasing complexity of design. Implications related not only to design but also to assessment, operation, maintenance and optimisation thanks to the use of real-world data in models and simulations.

Over the last few decades, there has been a progressive spread of sustainable design and technology solutions to ensure indoor climate control (ICC) in architecture. At the same time, a new cultural approach to design based on sustainability has spread among many operators in the sector. The current frontier of innovation in ICC lies in the development and application of passive and hybrid technologies and systems (Matteoli and Pagani 2010). Moreover, sustainability policies, including directives, laws and regulations, have, in recent years, been consistently implemented by the EU and by local and national governments, with an indirect effect on the growth of "environmental awareness" among end-users.

The applicability of passive and hybrid solutions for cooling and heating requires considerable attention to the local and climatic context as well as careful adjustments based on changing boundary conditions. This attention requires mature data, information and design and evaluation techniques suited to the management of complexity, also in view of the European directives on NZEB buildings (EPBD 2010/31/EU) and future net-positive buildings and their implementations—see, for example, the recent EU Directive 2018/844 of the European Parliament and the Council, 30 May 2018, amending Directive 2010/31/EU on the energy performance of buildings and Directive 2012/27/EU on energy efficiency (European Parliament and the Council 2018).

Data field and digital design | theoretical implications

The networking of nodes capable of collecting, communicating and transmitting data is becoming a tangible reality for the cities of the future, characterised by different approaches based on the different methodologies used. Proprietary platforms, like those developed for the Internet 3.0 and the Internet of things (IoT), are high-quality tools which do, however, pose some risks to data ownership and access, especially in a public–private environment.

The text is organised into practical applications and theoretical speeches. The former support theoretical argumentation, focusing particularly on three macrothemes:

- the scientific model as a design tool, in which they convert innovative technologies and IT tools for virtual, real or hybrid projects. Growing awareness of the potential of digital media to support architecture is changing the design process by accompanying simple representation with the addition of other specific technical and decision-making functions. The design possibilities that can be achieved with digital technologies are studied on the basis of the effects and the complexity and extent of the implications, using the concepts of I, II and III era/scope of digital application;
- platforms, management spaces and design. Platforms are the physical or virtual space for decision-making and process management (design, decision-making, organisational), where skills and techniques meet in specific ways. Platforms house the management of the different phases that follow the processing of data from production to use, exploiting modelling and materialisation techniques. This is why platforms are configured as places for managing complexity.
- data and its ownership. This topic focuses on the implications that different ways of producing, using and disseminating data and information have on cities and their inhabitants, linking collective or individual interests with the openness or protection of the systems used.

The study reveals the need to assess the implications that ICT and digital technologies in general have and will have on the social fabric, and on the urban and architectural fabric in their dual role of enabling and co-evolving, right from the earliest decision-making stages of implementation of strategic plans for smart cities.

The research presented in this book has been partially funded by Ministry of Education grant no. 48—B.s. 3rd (Youth Fund) "Parametric analysis of data from the monitoring of energy performances of *passive* and *hybrid* HVAC systems". In addition, the implementation of the new extended materials, including upgrading and English language adaptation, has been co-funded by the 59_RIL16CHG02 and the 59_ATEN_RSG16CHG grants.

References

European Parliament and the Council (2018) Directive (EU) 2018/844 of the European Parliament and the Council of 30 May 2018 amending Directive 2010/31/EU on the energy performance of buildings and Directive 2012/27/EU on energy efficiency. Offl J Eur Union, 61: 43–74.

Matteoli L, Pagani R (eds) (2010) Cityfutures: architecture design technology for the future of the cities. Hoepli, Milano.

Pfister C (2011) Getting started with the Internet of Things. O'Reilly, Sebastopol.

Wiener N (1950) The human use of human being. Houghton Mifflin, Boston.

Contents

Chapter 1
ICT, Data and Design Issues

To measure is to know.
If you can not measure, you can not improve it.
Lord Kelvin
The real revolution is not in the machines that process the data, but only in the data itself and in the way we use it.
Mayer-Schönberger and Cukier (2013)

This chapter briefly introduces some reflections on the repercussions that new information and communication technologies have on society and on design. Starting from a in-context reflection on the implications of digital tools on design, we analyse the concept of model and parameter. Below are some examples of parameters that are influencing and have influenced the forms, conformations and ways of living in a propaedeutic vision compatible with other fields of research such as parametric urban planning and the extensive use of geographic information systems. These parameters may not all be closely related to the digital world but linked to data and information and the way in which they are identified, analysed and defined. Lastly, the chapter introduces and examines the main areas of interaction and change between society and ICT.

As will be highlighted, data and information are subject to changes in the way science is organised and developed, so we can say that "we are reinventing discovery" (Nielsen 2012, 13).

Furthermore, the creation of a SmartCity, a strategy for an urban future on which the growing interests of the European Union and Italy are focused, includes, among other actions, a strong interest in the techniques of digitisation of the city (Chiesa 2013). The smart city, which by its very nature should be cohesive, connected and innovative, requires information and models that allow the measurement of order and disorder, or rather urban entropy, which, as recalled by Wiener in 1950 in *The Human Use of Human Beings*, is often cited as the scientific basis for a cautious vision of progress. The theme of entropy and the relative

> degree of a system can therefore also be defined in another language: that of information. In this language, the order of a system is equal to the amount of information necessary to describe it [...] but it is always legitimate to consider that one of the fundamental statements of the theory of information– i.e.: that the transmission of a message has to be associated

G. Chiesa, *Technological Paradigms and Digital Eras*,
PoliTO Springer Series, https://doi.org/10.1007/978-3-030-26199-3_1

with a certain dissipation of the information it contains - is the equivalent, in information technology, of the second principle of thermodynamics.

Monod (1970, 188)

The construction of the model based on what is real and materialisation (opposite procedure) are critical points for modelling in its varied applications and, as we are reminded by Minsky, models are simplifications-abstractions of what is real. A model can be a physical representation, a maquette, a simulation, created using cfd (Computational Fluid Dynamics) software, for example, but models can be human too. The term model, derived from the Latin *modellus*, is defined by the *Treccani Dictionary of Physical Sciences 2012* as:

- Construction that reproduces, usually on a reduced scale, a physical system [...] to study its behaviour in certain situations (m. in similarity);
- analogue reproduction of behaviour [...] of the system under study by a system governed by laws like those of the system under examination (m. analogue);
- by abstraction, system of equations that represents (m. theoretical real) or is assumed to represent (m. theoretical hypothetical), the laws governing said system, with parameters depending on the particular situation, and from which it is possible to deduce the system of equations or the equation that represents, within certain limits, the state of the system determined by given types of stress.

According to Jeff Rothenberg, "modelling is one of the fundamental processes of the human mind. [And a] model represents reality for the given purpose; the model is an abstraction of reality in the sense that it cannot represent all aspects of reality" (Rothenberg 1989, 75). Each model is characterised by three essential attributes, reference, purpose and cost-effectiveness, integrated with that recalled by Kiviniemi in a conversation held in 2012 in Turin. Three fundamental aspects of the model need to be considered:

- you need to define the goal(s) before you can build an appropriate model;
- different project areas require different models;
- the integrated model must include (at least) the parts necessary for the desired level of integration.

From a philosophical point of view there are different interpretations of the concept of model. Leaving aside the Deleutian idea, referred to indirectly in this chapter, we can look at the link that exists, in the Kantian vision, between model and representation. On the basis of this interpretation, the categories of intellect become schemes that allow us to codify reality in which the scheme is not a tangible representation of an object, but a rule for the construction of reality—see also (Chiesa and De Paoli 2014; Pagani and Chiesa 2016). However, if the model (real or virtual) has accompanied the history of architecture since the Renaissance (Maldonado 2007) and since the birth of the modern figure of the architect, in the form of physical models interposed between the ideas of architects and builders (Sass and Oxman 2006), it is today's modelling technologies that allow the management of contemporary design complexity. We are able to simulate reality with new levels of detail thanks to the increase in the power of calculation now reached by computers and the standardisation of models—e.g. CIM (Common Information Models). Examples of possible applications include thermal simulations of internal urban compartments, applying FEM (Finite Element Method) methodologies to models with of millions of surfaces, capable of finely simulating phenomena such as irradiation in urban compartments (e.g. Beckers 2012). And increasingly precise methods are being used to simulate and predict energy consumption on an urban or district scale, and production from renewable sources—see, for example, the works reported by (Bottaccioli et al. 2018, 2017)—and the extension of the study and simulations of urban scenarios from an urban metabolism perspective (e.g. Robinson 2011).

New methods of production and management of the link between project and material will, in fact, allow the management of high degrees of complexity and model-reality compliance in a context in which the level of complexity will be unrelated to the cost of labour. However, risks related to overwork, increased responsibility and lower profit margins, as in the case of the building information model (Bim) (Celento 2007), will need to be addressed. The progress of the big data revolution, in a perspective in which information itself, as well as technologies, will be subject to innovation and development, will be favoured by the extension of datasets from a small sample to the whole population (Mayer-Schönberger and Cukier 2013). These techniques and tools allow the holistic integration of specialists and different tools capable of supporting every phase of the life of constructions into a reference framework that is increasingly articulated and characterised by numerous needs.

Intervening on the urban fabric requires the increasingly multidisciplinary and complex organisation of work, involving the interaction of many skills in order to manage and implement successful policies of urban regeneration both in the design phase and in the maintenance phase. The construction and management of the resulting models are linked to the quantity and quality of the parameters (and data) that populate them. Moreover, the accessibility of information and the administration of models and actions of materialisation (actuators) pose very interesting ethical and practical questions. The impact of new information and communication technologies (ICT) on the social fabric and urban policies is also an issue to be analysed and planned at the same time as the creation of urban platforms for smart cities.

1.1 Background—Design, Technology and Information | IT with Emphasis on Technology

> Sometime during the 1980s the technological society which began in the fourteenth century came to an end. Now I recognize that dating epochs involves interpretation and perhaps some fuzziness in assigning beginnings and endings; but, nevertheless, it appears to me that the age of tools has now given way to the age of systems [...]. For most architects the word technology still means the different means and methods of building, however, in recent decades, the term has become synonymous with computers and the whole apparatus of networked information flow. With that change, the shift described by Ivan Illich in the opening

> quotation has become wholly palpable: architecture and technology or, more precisely, the tools of design and construction, have become a matter of systems.
>
> Braham and Hale (2007)

These words introduce the long list of essays that, based on the evolution in time of the conception of technology in architecture, introduces the advent of a new paradigm directly linked to the fate of the man-machine civilisation defined Ray Kurzweil in 2008 as *singularity*.

At the same time, for some decades now, contemporary architecture has been interfacing with new tools and possibilities interrelated with digital design techniques, rapid prototyping, automated construction and opportunities related to the parametric and algorithmic world. The innovation of tools and techniques makes it possible to dynamically intervene in the need-driven approach to design (Cavaglià et al. 1975; Bocco and Cavaglià 2008) radically changing the way systems and concepts of standards and regulations are managed (e.g. Braham and Hale 2007; Celento 2007; Oxman 2006; Mitchell 2005) (Fig. 1.1). See Chap. 4 for a detailed description of the need/performance-driven approach and its implications in the digital environment, with particular reference to the potential of explicit with respect to implicit design practice—see also Chiesa (2017a).

However, it is essential to take an integrated approach to the development of research and innovation in order to ensure the cohesion of specific and systemic implications (Chiesa 2010). On this subject we should heed the words of Ludovico Geymonat

> Is there an outline of vision in specialism or not? I believe there is; as I have tried to show in all my works, the specialist choice implies a vision of the world; specialism must, I would say, be reinterpreted not denied, and here the use of the word "dialectic" which denies and affirms is deeply analysed. Specialism is, in a certain sense, denied but also accepted, as the starting point for its overcoming; otherwise we have closure-specialism, which takes a reductive view even of issues inherent in itself.
>
> Geymonat (1986)

Meanwhile, although the question of the usefulness of computers in architectural design has been open since the 1950s, there are now two main approaches to the subject which, as Antoine Picon points out, are summarised by the terms

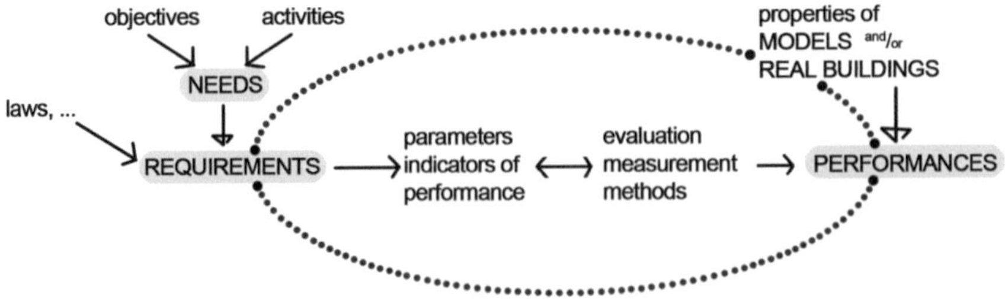

Fig. 1.1 Schematic interpretation of the requirement-driven approach, according to the definition in Bocco and Cavaglià (2008)

"*computerization*" and "*computation*" (Terzidis 2006). For many architects, the computer is an advanced tool for the operation of programs that allow the faster and more controllable reproduction of a design, the substance of which is consolidated and not modified by computer science. This is a level of use of digital media for representation only, i.e. the use of computer tools as electronic drawing instruments in accordance with a perception that still stems from paper-based methodologies (Kocaturk 2012). This category also includes most of the designers of blob architecture, although the articulation of their buildings, and their ideas, might make you think otherwise. The spread of the different ways of understanding design and using IT tools in the architectural field is a direct consequence of the fact that it is "unavoidable to enter into the black box of programming in order to make a truly creative use of the computer" (Terzidis 2006, viii). If the current use of computers in architecture involves a process based mainly on *computerization* (automation, mechanization, digitization, conversion) and, therefore, on the translation into the digital world of traditional logical processes, it is possible to use computer science for *computation* processes. The approach based on these processes consists in determining something using mathematical formulas or logical methods, exploring something indeterminate, often poorly defined, and subject to exploration. In theory, as long as a problem can be defined in logical terms, it is possible to find a solution to the questions posed by the question itself.

At least two levels of pervasiveness of computational techniques in digital design can be identified (Kocaturk 2012; Woodbury 2010). The first is based on the concept of the parameter, in which relational connections between elements vary in an output spectrum defined by an input domain according to a clear relationship f, where the elements are interdependent. The second level regards algorithms, typical of an application based on the development of the logic of the computational project, in which architecture is part of a flow of functions, of algebraic, geometric and analytical operations. Unlike the parametric concept, at algorithmic level the relationship between input and output is linked to the characteristics of the method of use and the formal description of the function f (Kocaturk 2012). In other words, "an algorithm is a process of addressing a problem in a finite number of steps" (Terzidis 2006). Terzidis also reminds us that an algorithm can, therefore, be an articulation of a strategic plan to solve a known problem or a stochastic search for possible solutions to a partially known problem or, again, a linguistic expression of a problem, composed of linguistic elements and graphemes, defined grammatical and syntactic rules. The same algorithm can give completely unexpected results when used according to parameters other than the original ones.

While the use of parametric models in architecture is a recent practice, unlike other disciplines related to the engineering context in which such practice began around the turn of the millennium, as Woodbury reminds us, the first cad system developed was parametric (see Ivan Sutherland's Ph.D. thesis on the Sketchpad system in 1963) (Sutherland 1963, 2003). The parametric approach is closely linked to the development of topological solutions, associative geometries, morphogenesis and animation of results, while the algorithm is more closely linked to an evolutionary approach, based on a systemic order capable of differentiating results and equipped with self-organizing techniques typical of generative models. An algorithmic model requires a topological order to be established and, in this sense, is a step forward compared to the parametric model (Woodbury 2010). This vision is compatible with the increasingly comprehensive application of what Oxman, in 2006, defined as "the digital age" in architectural design. The importance of this radical change had already been stressed as fundamental in David Celento's multifaceted article *Innovate or Perish*. As the title indicates, he states that "architects' refusal to embrace technological innovations invites their extinction" (Celento 2007). The current capacities of similar sectors, which are more open and capable of managing mass customization processes, to integrate design and production (in an evolution from *rapid prototyping* to *rapid fabrication*), to manage a parametric and algorithmic approach to design and to interface with the world of bim, led Martin Simpson of Arup Associates to suggest that "architects may eventually become unnecessary". In addition to this, it is useful to detect the resistance to the dissemination of increasingly small sensors and computers capable of surrounding and blending into the real world, and the difficulty of managing the concepts of brand and fame. The possibility of fully expressing the different digital eras—a concept derived from Oxman and related to the idea of application and its pervasiveness (Chiesa 2017a)—that are appearing, whose technological and informative components are evolving and consolidating, creates considerable difficulty for the traditional architectural world, which finds it hard to accept and grasp the transformations underway. Celento also says that "for architects, embracing these possibilities is on the one hand frighteningly simple and on the other hand scandalously improbable, for it involves nothing more than the abandonment of thousands of years of precedent" (Celento 2007, 8). We have to face and overcome those closure-specialisms which, consolidated by centuries of practice, end up, as in the case of big data, underestimating or completely missing the change. "It is a change that many scientists have not grasped or have underestimated, so absorbed by their specialist work that they fail to understand how vast the impact of new cognitive instruments is; they are like surfers, too busy watching the foam of the waves to notice that the tide is rising" (Nielsen 2012, 13).

1.1.1 Digital Design

> digital design is a methodologically unique form of design
>
> Oxman (2006, 238)

The concepts of standards and regulations are an integral part of the cultural logic behind the procedures of our way of understanding and dealing with design and design methodologies. However, the "logic of repetition [based on the concept that] the module is a formalism that can generate through reproduction; and reproduction, or repetition, of the elemental normative knowledge produced a world of normative order that was so fundamental to the age of machine industrialization" (Oxman 2006) is perpetuated today not with the aid of paper but using digital information (Mitchell

2005). Design is now a digitally mediated process that allows the creation of architectures dominated by complexity and, at the same time, capable of really meeting the contextual needs of the site and the requirements of the customer. The use of digital mediation implies a series of potentialities which, from the most conceptual and design-oriented phase, allow a design process strongly linked to local specifications and to the required performances. In fact, "if the new design is in any sense revolutionary, it is so not due to its forms, but to its ability to propose meaningful alternatives to the logic of repetition in the comprehensive historical sense proposed by Mitchell" (Oxman 2006). With regard to these statements, see, for example, the research and applications presented in the proceedings books of the *Advances in Architectural Geometry symposia* (e.g. Adriaenssens et al. 2016; Block et al. 2014; Hesselgren et al. 2012; Ceccato et al. 2010).

This opens up new ways of interaction between the real, dynamic and constantly changing world, and the designed world, also capable of becoming cohesive with a dynamic and evolutionary vision. In the 1990s, the role of digital design became a central theme in international theoretical debates, thanks to the spread of CAD, which went from the conceptual phase of the 1960s to the first applications of the 1980s (e.g. Penttilä 2006). Nevertheless, new design paradigms that hide a criticism of the formal complexity proposed by the previous vision of digital media for the project are now emerging (Oxman 2006). Beginning in the 1970s, the work of some scholars, including Mitchell and Negroponte, was already in line with this mature approach to the use of it in design. Then at the end of the 1990s, the concept of the 4-D model, capable of integrating time into the three-dimensional model according to an evolved view of life cycle techniques, arrived. More recently, we have seen the arrival of 5-D cad, with the introduction of cost, and there will be future evolutions from 2D cad to n-dimension cad (see, for example, Penttilä 2006; Lee et al. 2002). The development of models capable of integrating n dimensions stems from the consideration that "the space of information is by no means limited to the three dimensions. The expression of an idea or a chain of thoughts can be provided with a multidimensional set of pointers, which allow further processing or synthesis, which you can reference or ignore" (Negroponte 1995). On the basis of this evolution, the role of data and the focus on the information side of it will lead, in the near future, to new concepts and possibilities in design, with regard to the evaluation, visualization, simulation and quality of solutions, according to a perspective that can go from *how it looks* to *how it feels and sounds* (Penttilä 2006, 398). The digital design culture linked to the interaction of ICT in design must provide for a strong correspondence between theories and methodological and operational descriptions, to prevent complete loss of clarity. With the evolution of ict-society-design interaction axes, knowledge and direct mastery of the latter will become an essential prerequisite for understanding and formulating the theoretical asset related to their use (Oxman 2006). In fact, the digital and paper-based worlds are completely different. "In the digital world the medium is no longer the message. It's just a materialization of it. Starting with the same data, it is possible to have automatically different materializations of a message […] [while] shape will be conveyed to the data increasingly in the receiver and not in the transmitter" (Negroponte 1995). Negroponte also reminds us that the concept of interaction is implicit in the concept of multimedia. Digital design implies the active interaction of the designer, who will, therefore, have to know and understand digital environments and not only digitally generated representations or, worse, digital representations (Oxman 2006).

The transition from the industrial age to the digital age can be effectively summarised using the research carried out by Tuba Kocaturk in 2012. In the industrial era, mass production was a formal language for representing the technological framework, where technology is a tool applied to design within the scope of disciplinary operation of skills related to architecture. In the digital age, on the other hand, technology is an integral part of the design process, allowing a formal variety (cad, cam, cae, Mass Customization) in a multidisciplinary and collaborative context in which the different skills are compared and integrated. From the middle of the twentieth century, computerized control techniques began to allow the development of a mass-customized mentality, in which products, although industrially produced, are unique and not standardized as they were prior to digital design. Four main schematizations are identified from modernism to the contemporary digital age, working on the relationships between three figures and the areas linked to design, which are embodied in conception (architect), production and construction (Kocaturk 2012).

- At the beginning of the 20th century, modernism allowed the creation of a direct axis between the players involved in conceiving design and production. Le Corbusier with the metaphor of the machine (e.g. Corbusier 1923) and Grophius with the Bauhaus strengthened the birth of a production of building components based on standardization, mechanization, mass production and industrialization of the sector. The relationship between the designer and builders focused mainly on the concept of construction type and typology;
- in the Fifties, following the world wars, the modernist vision of design changed. First of all, architects gradually saw their influence on design diminish, creating a fracture with the nodes of production (now industrialized–manufacturing) and construction (now construction industry). The link between production and construction is a difficult area as it requires the combination of standardized production based on minimal variations and mass production with the construction industry, characterized by the need to deal with designs that are all different and whose performance is not easily predictable. The node of the construction industry is articulated as a system based on the contractor, on prefabrication, on the building considered as a system, on the division and specialization of roles and on systematization;
- at the end of the 20th century, the digital era model appeared, connected with the widespread adoption of cam and the birth of dad (Digital Architectural Design). In this scenario, conception is linked once more to production, according to a paradigm dictated by flexibility and mass customization, where industrial objects can be produced with customized and unique shapes. Design is characterized by a topological rather than typological approach, by a vision in terms of components rather than systems, by digital modelling rather than sketching. Designing in a digitally mediated environment allows the use of new strategies related to the

mental construction of the model and techniques of communication between the members of the design team, in a new perspective that unites designers, the subjects of the design and information. The criticality of the axis between the nodes linked to production and the construction industry occurs once again, due to the need to re-modulate that developed in previous scenarios;
- the 21st century sees the development of the themes of the digital age in the construction industry. The previous scenario is altered with the acquisition of new relationships between the node of design and that of the construction industry, and between the latter and production. The birth of overlapping and blurred disciplinary roles, the adoption of bim rather than paper-based technologies for documentation, and the use of targeted and specific productions as opposed to the systematization of components, opens up new scenarios in which the role of centralized management declines.

The four scenarios presented and described above allow an approach to digital architectural design that includes some peculiar characteristics. The transition to the algorithmic level of design will be resumed in the next point, introducing a detailed description of the first digital era, as theorized by Oxman. A further approach to the interaction between ICT and design in the digital age is reported in the Verb series, six volumes with a specific theme, on the conception of design in the digital age: the evolution of processes (Salazar et al. 2001), the context of formal and material possibilities in the information age (Ferré et al. 2003), connections in the world of interaction (Ferré et al. 2004), architectural identity in the age of real artificiality (Ferré et al. 2005), the artificial-natural relationship in morphogenesis and evolutionary models (Hwang et al. 2006), conflicts and crises between the limits of architectural design and the demand derived from balanced, dynamic and complex conditions (Ballesteros et al. 2008). In recent years, moreover, there has been a constant increase in the number of texts dedicated to the subject, as can be seen by simply searching the main databases of books and magazines—in ScienceDirect for example, the number of hits relating to the keyword "BIM" has grown from 149 in 2000–1355 in 2018—viewed December 2018. Furthermore, the study of DIM (District Information Model) systems is beginning to accompany the spread of BIMs —see for example (Osello et al. 2016) and the results of the European project DIMMER—in a perspective increasingly oriented towards integration between the different scales.

1.1.2 The First Digital Age

> It is significant that the centrality of the designer can be maintained in models of digital design. In fact, this concept has profound implications for digital design media in that it implies that the control of digital processes, complex as they may be, is based upon interaction and reflection with the designer.
>
> Oxman (2006, 241)

> Perhaps for the first time, through the invention of the computer, a device originally intended to serve people, ironically they were faced with phenomena that demarcated the limits of the human mind and cast some light into the corners of an alien world.
>
> Terzidis (2006, 55)

The definition of the first digital age as formulated by Oxman proposes the methodological framework of digital design in a vision that keeps the role of the designer at the centre of the process. However, while the designer remains at the heart of the process, the quality of his work is very much linked to "the nature of interactivity and type of control of design processes" (Oxman 2006). The role of the designer in the first digital age is, in fact, defined by the nature of his interaction with new tools. The architect is required to control, moderate and interact with the generation of geometries, with the algorithmic mechanisms of control and with the processes of management and optimization of performance. In this sense, the approach to design in the digital age requires constant ontological effort in the construction of models, as knowledge is rendered according to an explicit and not implicit perspective—see Chap. 4 and references in (Chiesa 2017a).

The sub-processes of digitally mediated design involve or may involve different types of interaction with the designer, including the possible sharing of models and knowledge—see also in this sense (Carrara et al. 2014).

In order to methodologically and theoretically describe the operational context of the first age of digital design, four sub-components of design borrowed from the traditional design practices yet radically different in terms of process and interaction are defined: representation, generation, assessment and performance. In traditional paper-based design, models are built on the basis of implicit knowledge (particularly in the generation and assessment phases); within the scope of the computational design, however, cognitive processes must be explained by the very nature of the computational process, as "parametric modelling (also known as constraint modelling) introduces a fundamental change: "marks", that is, parts of a design, relate and change together in a coordinated way. The act of relating requires explicit thinking about the kind of relation' (Woodbury 2010, 11). There are essentially two types of relationship between the sub-component designer and the different components: interaction and information flows (Woodbury 2010; Oxman 2006, 243). In the digital environment, the interaction that took place, in the case of paper-based design, directly between the designer and the forms he drew on the sheet of paper, depends on "specific implementations of computational constructs. Interaction with computational design media requires of the designer a different form of input and level of formalization" (Oxman 2006). The difference between paper-based and digitally-mediated design models is substantial and acts on both a theoretical and a cognitive level. Interactions can be external, i.e. traditional direct interactions with forms and geometries, and internal, in cases of interaction with forms mediated by the digital medium, the digital environment, the computational process or the algorithmic mechanism (Kocaturk 2012; Oxman 2006; Terzidis 2006). Oxman introduces four main classes of interaction:

- interaction with a free form (paper-based and non-digital representation);
- interaction with digital constructs (typical representation of CAD drawings where the designer interacts with a digital drawing or model);
- interaction with a digital representation generated by a mechanism (typical of interactions with mechanisms of generative design, the interaction of the designer occurs with a digital structure built on a mechanism based on predefined rules and relationships—parametric approach);
- interaction with digital environments that generate a digital representation (typical interaction with the operational components of the mechanisms of the generative project, the interaction of the designer takes place on the computational mechanisms that generate digital representations—algorithmic approach).

Based on what has been introduced, we describe some of the schemes developed by Oxman aimed at defining different conceptual and methodological models of design. Processes and interactions between the designer, form generation, assessment and performance occur implicitly using the architect's cognitive framework. The interaction between the designer and the formal representation takes place directly, in compliance with the free form model.

The author uses an approach based on the modelling process to identify five models of the possible interactions (and degree of interaction) of digital design processes. The description of the models is deliberately forced into a scheme of five squares, which can be arranged in the shape of a cross, in order to facilitate the reading of the differences on the basis of a matrix compatible with traditional design. The designer is at the centre of the cross, while the representation, the generation of geometries, the performance and the assessment are at the four ends.

The model based on the traditional use of cad is divided into two main sub-models: the descriptive use of cad and the generative-assessment and predictive use of cad. The descriptive approach reports the interactions between 2d and 3d cad models developed after the design decisions have been made, in compliance with a process in which cad tools are used as substitutes for traditional manual design with some small qualitative effects. The predictive use of cad assesses the model, implicitly generated by the designer, using a digital process according to some macro-categories of indicators, economic, structural and environmental for instance, making the links between the designer, the shape assessment process and digital representation, and the link between the latter and assessment, explicit. These functions are destined, however, to change radically with the interaction between the digital model and its materialization into physical models, thanks to the techniques of integration between the real world and the virtual world (e.g. Sass and Oxman 2006). This process is not only an evolution of the use of cad for the description and/or assessment of design (e.g. Kocaturk 2012), it has implications of such importance that we can say that we are at the dawn of a new industrial revolution (Anderson 2012) in a more mature approach that can be defined as the second digital age (Chiesa and Casetta 2012).

The model based on the digital formation of models, in which the concept of shape changes into formation, in a dynamic conception of the role of representation. "The form generation in this model is based on interaction with an enabling digital technique […] rather than with the explicit representational structure as in the CAD model" (Oxman 2006, 250). In this sense, the designer, as the user of a technology, becomes the creator of specific tools for design, encouraging the architect to get in touch with programming (Oxman 2006; Terzidis 2006). Oxman identifies three main sub-models:

- topological design (for the construction of digital media for *forming,* like hyper-geometries, hypersurfaces, blob architecture;
- associative design, based on the principles of parametric design for the construction of associative geometries, using software such as Catia and Generative Components);
- dynamic design, focused on *morphing* and animation, in a dynamic approach to forming where the design is developed using "an interactive process of parametric change" like some uses of Kangaroo in Grasshopper (Lynn 1999).

The model based on generative design has a significant, albeit subtle, difference from the previous model. In fact, "here, as compared to formation models, the designer interacts with the generative mechanism" (Oxman 2006). Shapes and geometries are the result of a generative process formulated upstream in a context based on interaction. Generative design presents a decidedly vast and articulated cultural context that can be mainly divided into two sub-models: *Shape grammar* (e.g. Hou and Stouffs 2018; Ruiz-Montiel et al. 2014; Ceccato et al. 2010; Pancini et al. 2010)—also linked to the first studies on performance-driven architectural design in Italy (e.g. Cavaglià et al. 1975) and its impact on technological-environmental design (e.g. Chiesa and Grosso 2019; 2017; Grosso 2005)—and evolutionary models, based on the natural generation of shapes (Eilouti 2018; Chiesa 2017b; Chiesa 2010; Mateos 2007; Dollens 2006; Hwang et al. 2006; Mitchell 1997, 1998). Remember Xfrog software and genetic codes for forming and/or optimization, like Galapagos for Grasshopper.

The design-based model capable of generating a shape from specific performances. "The object is generated by simulating its performance" (Oxman 2006, 257). Two sub-models can be identified: the performance-based forming process and the design of performance-based generative models. In the first case, digital simulation of external forcings is used to control a shape forming process. For example, Foster and Partners' Swiss RE was developed on the basis of a forming process led by a performance-based approach (Oxman 2006; Hwang et al. 2006, 50–53). In the second case, a generative process is led by a performance-based approach integrated with the definition of further processes. This sub-model is connected with the next step based on the enabling integration of digital design media. The performance-based generative model illustrates a design process in which data and information are derived from simulations of the performance of geometric variables and do not guide the shape generating processes. The designer can interact in the three axes identified: defining the performance model and simulations, acting on algorithms and generative parameters and acting on digital representations (Oxman 2006, 259).

The model based on a composite and integrated approach can be defined as "a class of future paradigmatic digital design media […] based on integrated processes including formation, generation, evaluation and performance (this model allows different modes of interaction for each node with data and information flows as a) compound integrated network of enabling design media" (Oxman 2006, 260).

The increased complexity and potential of digital media for the project leads to the emergence of specialized professionals capable of designing using numerous types of software, developing programs and scripting, manipulating and updating increasingly complex and articulated models and datasets rich in information (Mayer-Schönberger and Cukier 2013; Celento 2007; Oxman 2006). An example is the parametric systems, of which BIMs are just a small part, and the specific skills needed to use them consciously. In a context in which elements can be "determined solely by the automated definition process, no condition is individually designed. The effect is not random or accidental, but it is locally uncontrolled—determined only by the global decision on the rules" (Sakamoto and Ferré 2008, 48). Terzidis's statement "the problem with this situation is that designers do not take advantage of the computational power of the computer" (Terzidis 2006, 58), will be surpassed by the birth of a class of writers of digital systems, defined by Oxman as *digerati*.

1.1.3 The Second Digital Age

> It's just that the digital revolution has now reached the workshop, the lair of Real Stuffs, and there it may have its greatest impact yet […]. From a design perspective, this is revolutionary. It is no longer necessary for the designer to care or know about the manufacturing process, because the computer-controlled machines figure that stuff out from themselves. The same design can be fabricated in metal, plastic, cardboard, or cake icing.
>
> Anderson (2012, 13, 86)

The conceptualisation of the first digital age is centred on the digital design process, focusing mainly on modelling techniques. Digitally mediated and created models are not only virtual, real or hybrid objects, but also places of integration and interaction. Along with the enrichment, complexification and articulation of scientific modelling techniques for design, reification and materialisation methods that involve a modification or change in the theoretical construct and digital tools for the designer are being developed. As mentioned with regard to the CAD-based, along with the visualisation techniques of the first digital age, completely new modes capable of materialising real objects from digital environments are spreading. This innovation, highlighted by Oxman himself, is an element of change with respect to the techniques of the first age, flanking them on a different front in terms of innovation, allowing us to speak of a real industrial revolution (Anderson 2012). This is an innovation in hardware, different from the first digital era, based mainly on software innovations, which is progressively investing in the processes of design and implementation of architecture as well as other sectors (Anderson 2006). This second digital age will only be able to fully assert itself following the enrichment and full application of the potentialities enunciated for the first age and inherent in the modelling process centred on the digitally mediated re-reading of the model. While modelling concerns the process of building the model, materialisation works in the opposite direction and deals with the passage from model to real object. This process is becoming increasingly evident in the digital world, where opposite processes of materialisation of the virtual appear alongside the processes of virtualisation—consider the business organisations that deal with additive manufacturing or consulting in the field, from three-dimensional scanning, to the optimisation of geometries and through to the creation of the printing file. Moreover, along with the technical possibilities offered by this change, there is another essential axis of innovation concerning "who's doing it [...]. Once things can be done on regular computers, they can be done by anyone. And that's exactly what we're seeing happen now in manufacturing" (Anderson 2012, 18). This is one of the premises of the second digital age, an age based on hardware innovation in which, alongside the demand for reality in modelling, there is the possibility to overcome it with the creation of real objects.

This not only allows the alteration of the modes of representation identified by Oxman, but also the entire process of constructing a building. According to Anderson's research, the revolution mainly concerns those who can access materialisation, in a perspective of widening the productive technologies of the maker movement. Evidently, makers will play a new role in the contemporary world, both in terms of widespread production and the development of shared solutions, such as Linux-style platforms in the real world. We must, however, remember that this change will have very strong repercussions on the industrial world and on design, starting with the possibility of shifting production, or part of it, from mass production to mass customisation (Kocaturk 2012), in which the same numerically controlled machines will go from producing thousands of elements that are all the same to thousands of elements that are each different from each other, to suit the requirements of users. The use of parametric patterns and faults extends from the world of modelling to the world of materialising custom-made objects through parametric and algorithmic design. The increasingly complex digital process will make it possible to materialise objects which, as in the case of the photographs of the film smoke, "are all the same, but each is different from the other", in that they show the same corner but at different times and with different people. The field of materialisation concerns a wide variety of possible applications, which, in addition to the world of makers, are oriented to 2.0 craftsmanship and overcoming traditional industry towards a 4.0 vision, the social and other impacts of which are far from being fully understood and solved. See also the interview with Stephen Hawking that appeared in the Guardian in 2014 (Clark 2014).

Hardware-based innovation will enable the digital design of objects and architectures that can be effectively optimised and localised using the tools of the first digital age, in the knowledge that they can then be produced and used without cost increases. The development of specific calculation algorithms, which are the subject of extensive international research (e.g. Ceccato et al. 2010; Sakamoto and Ferré 2008), allows the production of customised elements, minimising both waste and scrap. Production can also be no miles, considerably reducing the transportation of industrialised elements, from the transport of atoms to the movement of bits. An interesting example is the Local Motor Factory in Chandler, Arizona, which can produce fully customised cars. This is made possible thanks to the community of the project which, exploiting cognitive surplus (Shirky 2010) and dark energy (Anderson 2012) present on the net and in the members of the community itself, develops new objects in a short time, solving technical problems free of charge and optimising the products according to specific requirements. "Open innovation communities connect latent supply […] with latent demand" (Anderson 2012), stresses how the revolution of atoms in the world makes use of experiences and new paths of innovation in the fields of skills and science (Nielsen 2012).

From a technical point of view, numerical control machines are a technology that has been mature for several years and they make it possible to overcome the geometric limits imposed by the use of milling cutters, in right angle machining for example, thanks to the increase in the axes of freedom of CNC, as emerged in a conversation in Rome with Felice Ragazzo in 2010. In addition to this, domestic machines for the production of objects and components are being joined by machines capable of processing any material in any size. Behrokh Khoshnevis, of the University of Southern California, has developed a technology that demonstrates how innovation in numerical control production can also reach the construction sector, radically changing it (Khoshnevis 2004; Zhang and Khoshnevis 2013). A similar result was also presented by an English-speaking research group (Lim et al. 2012). Contour Crafting and other similar technologies are processes for the production of buildings and architectural components using special 3D printers, suitable for the use of concrete and other building materials on a large dimensional and quantitative scale (Chiesa and Casetta 2012). These techniques, which influence not only the production of the structure but also the installation of finishes, as in the case of some robotic arms (Khoshnevis 2004), will make the thought of Bass, according to which "we can separate the design of a product from its manufacture for the first time in history, because all of the information necessary to print that object is built into the design" (Anderson 2012, 86), effective also for the construction sector. As identified by the *designtoproduction* group (Sakamoto and Ferré 2008, 160–163), the process that connects architectural design, understood as defined in the first digital era, and numerical control production, understood as the overcoming of traditional cams, can be divided into four main phases:

- the organisation of the design process in a structured and parametric three-dimensional vision;
- the algorithmic optimisation of the model based on the interactions between the parties;
- simplification of the parts by rationalising them according to production;
- the materialisation of information through the production of data to produce parts.

The development of the second digital age is still in progress, but it is possible to identify some fundamental steps capable of describing a first articulation organised on the basis of the possible implications. The use of the processes of materialisation can, in fact, affect representation only, enriching the use of cad described above. In addition to representation, it is, however, possible to use materialisation tools to subject families of objects to real-world testing through the interaction between simulation and monitoring, partly anticipating what will be described as the third digital age. Another type of potential is the process of on-site production and distribution of production optimisations from a diy perspective, in which the distribution network switches from atoms to bits (see for example Anderson 2012; Negroponte 1995). We must remember that the maker movement is based on the principle of "give away the bits and sell the atoms" (Anderson 2012).

The second digital age, as a whole, allows us to overcome several critical issues related to the complexity of shapes and geometries typical of the new potential of the first age. As Anderson points out:

- differentiation is free. Differentiating objects costs the same as if you make them all the same;
- complexity is free. Making objects more complex and articulated does not increase the costs because digital modelling environments can easily elaborate complex and articulated algorithms;
- flexibility is free. Changing or differentiating a product already in the production phase means simply altering the code sent to the machine and not the machine itself.

The implications of this change can, on the one hand, innovate the movements aimed at designing local resilience in terms of creating local renewable districts (Chamberlin 2009; Hopkins 2008; Todd and Todd 1993), and, on the other, influence research in the field of hybrids between the natural and artificial worlds, printing cells and creating objects capable of self-assembly using local resources (Bar Cohen 2006). The *Silkworm* project (Holloway et al. 2012), for example, studies how produce geometries inspired by spiders' webs and silk threads using 3D table-top plastic printers, in order to overcome the classic use, in overlapping layers, of this materialisation technology. In addition to these two directions, it will be possible to influence the conformations of urban *smartness* by interacting with objects in what will be the web 3.0, i.e. the internet of things.

1.2 Parameters, Civilisation and Settlement Shapes | IT with Emphasis on Information

Settlement processes are influenced by numerous parameters and phenomena that determine and configure specific geometric, organisational, localisation and social variables according to complex and generally difficult to identify reciprocal logics and interactions. Individual design actions are dictated by aesthetic, functional, technological, need-based, cultural, social, political and budgetary variables, among numerous others, which make design a complex and articulate profession. However, in addition to these variables, the organisation of society and civilisation, with repercussions in terms of long-term needs, forms of living and functional specifications, is affected by other more complex and far-reaching parameters.

This paragraph identifies some of the macro-categories of parameters capable of influencing social organisation, urban planning and geometric forms of civilisation on long or medium-long timescales. The intention is not to provide thorough coverage of the subject but to look at certain instances, referring to specialised research and future insights for a more articulated view.

1.2.1 Macro-Categories of Factors Influencing Civilisation in the Long-Term

The various macro-categories of parameters that influence and have influenced forms of living in different time horizons include:

– the political and cultural organisation of society, such as the effects of feudalism or of the latifundium or the castramentatio romana (Chiesa and Di Gioia 2011; Vitta 2008; Magnaghi 1994);
– economic organisation with repercussions on landscape forms, land use, population distribution, trade and work at different scales (Sassen 2010, 2018; partially Turri 1998; Rhomer 1964);
– energy carriers used which affect the road network, energy networks, technological, architectural and urban choices (Droege 2006; Butera 2004; Grosso 1998; Hawkes et al. 1987);
– the climatic repercussions from which the location of fertile soils, ice and currents derives, with due repercussions on settlement capacity and on the type of social and urban organisation—see in particular what has been described by paleoclimatology scholars (Fagan 2002, 2009; Coppens 2006; Weiss et al. 1993).

It is also necessary to take into account the fact that currently one of the major geological forces capable of generating direct impacts on the biosphere is represented by humanity itself, which is why it is possible to identify, for expansion of the cultural dimension mentioned above, a further parameter capable of influencing civilisation in the long term:

– the anthropisation of the biosphere, which corresponds with the anthropocenisation of space (Foley et al. 2013; Slaughter 2012; Syvitski 2012; Crutzen 2005; Crutzen and Stoermer 2000).

Lastly, a recent parameter, the effects of which are occurring right now and are hard to predict, has been identified:

– the methods of production, transmission, collection and use of data and information on different scales, as well as the role and type of information technologies used (Floridi 2013, 2017; Occelli and Staricco 2002; Cohen and Nijkamp 2002).

This last parameter, based on it, is configured according to different action vectors, as described in Sect. 1.3, and can be divided into at least two main directions of influence: the first, due to the technologies component and focused on digitisation, the second on the information component and based on data collection. Moreover, the influence of data, information and related technologies extends to the modalities of scientific discovery (Nielsen 2012), where science is understood as a parameter capable of radically modifying forms of living and modalities of civilisation. The idea that it has specific effects on the way in which urban forms are developed pervades the analyses in this paper. Monitoring has been a way of acquiring information and building knowledge since the birth of modern science. Currently, with the progressive revolution of big data, the possibilities of data processing are growing exponentially, changing the operational scenarios. For example, we can remember how the use of big data can encourage the analysis and study of parameters related to the prosperity of cities and individual districts or to urban development or, again, to the spread of particular services and so on (see for example the case of the start-up *Eagle*, as mentioned by Mayer-Schönberger and Cukier 2013). This book is based on the recognition of the existence of this new parameter, although it is often configured as a set of tools for the management of other parameters of influence. However, at a time when the ways of understanding scientific research (Nielsen 2012), the big data revolution (Mayer-Schönberger and Cukier 2013) and the other repercussions described below (Floridi 2013) are changing require methodological and conceptual redefinition, it is hard not to see the repercussions, some of which formal, that ICT have directly on civilisation as previously introduced in Sect. 1.1.

1.2.2 Factors Influencing Urban Shapes (Medium-Short Time Spam)

In addition to the macro-parameters described above, below is a list of non-geometric determinants that influence the forms and development of settlements and which, at least in part, interface with the urban sciences and planning, as identified in a document of the *City Form Lab* of Singapore (Sevtsuk and Amindarbari 2012, 61). The determinants identified capable of influencing the geometric shapes of a metropolis are:

A. Economic-related determinants

   a. Average family income
   b. Gross Domestic Product
   c. Gini coefficient (income distribution and inequality)
   d. Crime rate
   e. Average household size
   f. Cost of living
   g. Land cost
   h. Number of employees
   i. Number of companies

B. Transport-related determinants

   a. Kilometer travelled per vehicle
   b. Transit of public transport
   c. Number of cars

C. Energy-related determinants

   a. Average consumption per household
   b. Energy cost
   c. Cost of transport energy

D. Policy-related determinants

   a. Regulatory system
   b. Corruption

E. Geography-related determinants

   a. Temperatures
   b. Water resources

F. Demography-related determinants

   a. Total population
   b. Percentage of elderly people
   c. Percentage of immigrants

The same document reminds us that the factors identified here are not exhaustive and serve mainly to stress the need to develop more in-depth studies on this subject.

It is clear that some of the parameters previously identified are missing from the list presented. No mention is made of the implications that act on broader time frames. This list could be interpreted as a list of determinants that act on design, while the previous list is related to parameters that affect civilisation in general. The fields of application and influence, as well as the time frame, are therefore different. There have been studies on the relationship between forms and urban energy consumption (e.g. Yang 2015), on the relationship between urban design and climatology (e.g. Erell et al. 2015), and on the geo-climatic applicability of sustainable technologies (e.g. Chiesa 2019). See also the studies on Urban Microclimate simulation developed by MIT—e.g. Urban design software and Urban Weather Generator (e.g. Nakano et al. 2015; Nakano 2015).

1.2.3 An Example: Climate and Civilisation—a Long-Term Factor

> As Stephen J. Gould, a biologist at Harvard University, once pointed out, we are all the product of the same African lineage; we share great reserves of human potential which, whether we are Native Americans, Europeans, Australians or Eurasians, made us react in a similar way to the whims of climate change during this long summer.
>
> Fagan (2009)

Climate conditions are a parameter that has undoubtedly influenced the history of civilisation. In fact "climate is [...] one of the fundamental dynamic scenarios of human experience" (Fagan 2009) and its effects throughout history can be traced in specific examples (Fagan 2002, 2009). The climate-humanity relationship is articulated around the concept of vulnerability to change on different timescales.

Civilisation and climate change

In recent decades, following the improvement of the techniques used to study climatic phenomena and the compilation of a larger amount of data to ensure a better time map than in the past, paleoclimatology is starting to be considered an important science for the study of the implications between climate and civilisation. The creation of models capable of simulating climatic phenomena allows a better understanding of these interactions (Broecker 1997)—see, for example, the different future scenarios reported in IPCC reports (e.g. the Representative Concentration Pathways—RCPs scenarios of AR5) and the new high-resolution climate models capable of simulating phenomena such as heat waves with resolution scales that can reach up to 25 km (e.g.: van Haren et al. 2015). A comparison of different high-resolution models can be found in (Haarsma et al. 2016). However, different models may provide divergent solutions, stressing the importance of comparing different results to measure uncertainty (e.g. Palazzi et al. 2017). These tools allow the performance of statistical hazard analyses to study the implications of climate change in certain territories (e.g. Alvioli et al. 2018), considering different parameters and analysing the differences induced by the original databases (e.g. Palazzi et al. 2013).

While climatic dynamics have not directly driven the events of human history, they have had numerous implications on adaptations and have influenced the organisation of civilisation as a whole. Remember that the Holocene is "the longest period of stable and warm climate that has been seen on our planet for 15,000 years ago" (Fagan 2009).

Starting from the birth of the genus Homo, the following are some of the climatic dynamics that have affected the most recent history of civilisation since 15,000 BC. Particular reference is made to changes that have taken place in the European area as a result of the dynamics of the Gulf Stream in the period between 16,000 and 3000 BC and to the subsequent consequences of the phenomenon known as *El Niňo*. These descriptions allow us to introduce the environmental vulnerability of settlements, which is addressed in the next paragraph.

The genus *Homo*.

Many the wonders but nothing walks stronger than man. This thing crosses the sea in the winter's storm, making his path through the roaring waves. And she, the greatest of gods, the earth – ageless she is, and unwearied- he wears her away as the ploughs go up and down from year to year and his mules turn up the soil. Clever beyond all dreams the inventive craft that he has which may drive him one time or another to well or ill.

Sophocles, Antigone, Chorus—about 442 BC

Since the appearance of the genus *Homo*, which occurred between 3 and 2.5 million years ago, the interactions between human history and climate dynamics have been evident. In fact, the climate crisis that hit the Omo Valley, currently between Ethiopia, Sudan and Kenya, between 4 million and 1 million years ago had repercussions on local species in terms of extinction, migration and adaptation. The adaptive responses to climate change developed by the hominids consisted in mainly in two solutions: on the one hand, the robust Australopithecus, and, on the other hand, the appearance of the genus Homo which, unlike the other, can be described as an *intellectual deterrent*. It adapted to the changed climatic conditions with an increase in the size of the brain and change of diet, switching from a vegetarian diet to an omnivorous diet (Coppens 2006). At least until the onset of the history of artificial things, the genus Homo was directly interrelated to and an integral part of the ecosystem and consequently endured changes in climate and environmental conditions just like every other living species. As far as the furthering of the themes related to the history of artificial things is concerned, reference is made to the research *Technology, science and history* (Cardwell 1972), remembering, however, as above in the Chorus of the Antigone (Sophocles about 442 BC), that it is reminder that " Many the wonders but nothing walks stronger than man." because man can pass through furious storms, can work with horses, he rules over beasts, and "he placed the yoke", but there is also the fact that man can operate both in the right direction, and in the wrong one, becoming an "arrogant man" perturbing nature.

The Gulf Stream.

Around 15,000 years ago, two global climate changes led to the reactivation of the Gulf Stream, which had "flowed at only two-thirds of its current speed" during the Ice Age (Fagan 2009; Lynch-Stieglitz et al. 1999). The atmosphere warmed up rapidly, while the level of the cold and salty water in the Labrador Sea rose considerably. This climate change in the atmosphere-ocean system changed the European climate of that historical period so much so that, after two millennia of rapid changes, the Europe of 13,000 BC was covered with forests. At the same time, the migration and extinction of a significant number of animal species pushed the *Cro-Magnons* to change their diet in order to cope with the different resources available (Fagan 2009). The diet of the Ice Age, mainly based on a carnivorous diet, was converted to vegetarianism. Changed diet and climatic conditions had a significant influence on lifestyle, not so much in terms of major social changes, but rather in terms of organisation in relation to the territory. The population began to spread over larger areas and to maintain the settlements for longer periods to suit their lifestyle as gatherers, while retaining the constant mobility of the hunter-gatherers.

A few millennia later, however, some populations, particularly those located in the territories of present-day Lebanon and Israel, substantially change the organisation and rhythms of society to suit periods in which they harvested their food crops, which became one of the determining factors in life. The need to wait for the various food crops to ripen determined the appearance of increasingly stable settlements, particularly on ecotones, i.e. the boundaries between ecological zones. However, the spread of more stable settlements, organised according to the natural harvesting cycles, was highly dependent on the stability of local climatic conditions. For example, the prolonged drought that hit Europe in around 11,000 BC led to the collapse of these first settlements, including Abu Hureyra (Moore 2000) in the Jordan Valley, which was abandoned after a first attempt at cereal cultivation.

In around 11,000 BC, in fact, the Gulf Stream, the main cause of climate change in Europe in previous millennia, was significantly reduced following fresh water flooding from Lake Agassis to the Labrador Sea. The drop in temperature recorded in Scandinavia has caused the advancement of the polar icecaps and the blockage of the Gulf Stream "favouring a much colder climate in Europe" (Fagan 2009). This interruption, known as Younger Dryas, lasted about a millennium and when the Gulf Stream resumed, the climate in Europe became mild again. In the period between the Dryas Younger and 8000 BC, new settlements based on agriculture and the introduction of animal husbandry were established. The search for reliable sources of food led to the massive development of agriculture and animal husbandry in large and permanent settlements. For example, around 9500 BC in Abu Hureyra there was a new settlement, bigger than the previous one and characterised by the presence of houses made of mud bricks.

In 6200 BC the Gulf Stream was blocked again, due to fresh water and the melting of the glaciers on the other side of the ocean, with the collapse of the Laurentide ice sheet. For about four centuries, we endured a global mini Ice Age which caused prolonged drought in South-West Asia. In 5800 BC, the Current began flowing again and consequently, for several millennia, Europe was characterised by a mild climate. During this period, farmers repopulated the fertile Crescent land, where the territory was particularly sensitive to climatic phenomena such as the attenuation of the Gulf Stream or changes in the course of the Indian Ocean monsoons due to El Niño events. The small agricultural settlements began to group together in large towns because "no one family could cultivate the floodplains on its own; everything depended on […] well-organised work teams, which worked for the common good" (Fagan 2009). The geographical conformation required the management of a network of embankments and canals capable of connecting watercourses with cultivated areas in order to feed a population that was experiencing a significant population increase. The big town of Uruk, for example, had strong commercial ties and a network of satellite villages that depended on one another in terms of crops that extended across as for a diameter of ten kilometres from the centre. The persistence of favourable climatic conditions allowed the formation and spread of settlements based on agriculture in geographical areas which had been mainly occupied by hunter-gatherer tribes until then. As of 3800 BC, the climate became drier and, in order to cope with periods of drought, the towns progressively transformed themselves into city-states that contended for the vital territories for the canalisation of the waters. "The Mesopotamian city was the human response to an environmental crisis" (Fagan 2009) and was capable of centrally organising the management of canalisation and the movement of people to larger towns. Between 2200 BC and 1900 BC, the climate changed further following a volcanic eruption and the beginning of a new drought that lasted almost three centuries, causing the abandonment and collapse of the cities. An article reported in Science in 1993 estimates that there was an exodus of 14,000–28,000 people from the city of Tell Leilan alone (Weiss et al. 1993).

A specific case: the city of Ur between the 6th and 3rd centuries BC.

The consequences of climate change on civilisation in the Mesopotamian area from the fifth to the third millennium BC, studied in detail in recent decades (Ur 2010, 412–413; Fagan 2009; Weiss et al. 1993), can be emblematically exemplified by the case study of the city of Ur, located in southern Mesopotamia. The first agricultural villages were already present in the area as far back as 6000 BC, when there was abundant rainfall. These first nuclei, built around proto-systems of water canalisation, consisted of huts built mainly with reeds and were similar to those found in other territories of Mesopotamia, both in the South, as in the case of Uruk (Fagan 2009), and in the North (Ur 2010; Bagg 2000). The plentiful food supply allowed rapid growth of the population with the consequent formation of first proto-city, capable of controlling its agricultural territory and managing the surplus in order to ensure survival for short periods of drought. Around 3800 BC, however, the monsoon of the Indian Ocean changed its annual route, causing a drastic reduction in rainfall with the consequent need to ensure livelihood by relying on other sources of water, including the channelling of river routes. The sowing period was altered to allow the exploitation of the river flood season (Fagan 2009; Aqrawi 2001, 273). This change led to the development of a new territorial conformation based on the increase in the rate of urbanisation and the foundation of large cities around the places from which they managed the water distribution channels (Fagan 2009). Along with the increase in urbanisation, social complexity, the centralisation of organisation and economic specialisation grew too (Ur 2010). These phenomena mainly affected cities that were growing rapidly both in terms of population, exceeding even 5000 inhabitants, and in terms of territorial extension, expanding control over the territory for radii of more than ten kilometres (Fagan 2009). For example, in around 2300 BC, the city of Ur rebelled against the rule of the older Uruk, located further north, and, under the leadership of the Sumerian king Ur-Nammu, began to extend its sphere of influence from the Syrian desert to the lands of the eastern Mediterranean. The city of Ur fortified its central core, with the construction of an impressive ziggurat, and outside the central walls, the territory underwent rapid urban renewal (Fagan 2009). Although Ur was able to withstand sudden periods of drought thanks to its careful storage of food surpluses, the city was totally dependent on river floods. Ur, like the other major Sumerian cities, was therefore better able to withstand small perturbative climatic phenomena than the original small villages (Fagan 2009). In 2200 BC, however, there was a sudden change in climate due to a major volcanic eruption (Ur 2010; Fagan 2009; Weiss et al. 1993) followed by a long period of drought lasting about three centuries. The consequences of this change in the flooding of the Rivers Tigris and Euphrates were evident: the lack of snow and the scarcity of water reduced the plains of Habur, which had been fertile and irrigable until then, to a semi-desert state. In the same period, periodic invasions by nomadic peoples intensified, making it necessary to build new defensive walls, like the *wall of Amurru* (Fagan 2009), and the population living within the cities increased. The consequences of this situation led to the collapse of the urban system which had previously thrived, so that "in 2000 BC less than 50% of the Sumerians still lived in cities". (Fagan 2009, 9). Only in the following century, after rainfall had resumed, was there a new cycle of urbanisation.

The example of the collapse of the city of Ur is emblematic of the influence that climate can have on civilisation, because, as Fagan reminds us, "for the first time, an entire city had disintegrated in the face of an environmental disaster" (Fagan 2009, 9). The urban organisation of Ur allowed the city to withstand small and recurrent climate changes more effectively than villages, but was not resilient enough to cope with prolonged droughts; in other words, the collapse of Ur occurred when the vulnerability threshold was exceeded.

The European ecotone.

Between 1200 BC and 900 AD, the European climate was affected by significant climate change. In particular, the ecotone between the Mediterranean and the Atlantic-continental climate underwent a first shift northwards around 300 BC, followed by a southerly adjustment around 500 AD, resulting in a drier and colder climate throughout northern Europe.

The various changes in the European ecological zones favoured different ways of land management from time to time, influencing the expansion of specific populations capable of being less vulnerable to the meteorological phenomena of the dominant climate. For example, "when the ecotone moved abruptly northwards, Rome quickly gained power" (Fagan 2009). During the expansion of the Mediterranean climate, the high resilience of the Roman Empire allowed it to withstand colder years and short periods of drought. However, this was not enough to easily guarantee control of the territories of northern Europe if the ecotone moved back southwards, causing prolonged periods of famine. From the third century BC, the Roman political crisis was accompanied by the start of a change in the European climate, bringing the ecotone close to the African coasts in the space of a few centuries. This climate change compromised the quality of crops on the one hand and caused the local people to return to a subsistence economy, accompanied by an increase in wariness and mortality, on the other. This condition changed only in 900 AD with a consequent new European agricultural development stimulated by the warmer and wetter climate.

Vulnerability, climate and civilisation

Throughout history, mankind has significantly increased its resistance to the most recurrent climatic events, through control and change of the territory, but has also significantly reduced its threshold of vulnerability to exceptional events and to long and medium-term climate change (Fagan 2009). The cost and the capacity to react to such events is in fact enormous and hard to tackle. Humanity has now *colonised* the entire planet, exposing itself, during the long period of climate stability that has lasted for 15,000 years, to the risks arising from any climate change. It is hard to predict the ability to easily manage phenomena such as mass exodus or profound geographical changes, which would occur in the event of a shift in these stable conditions (Fagan 2009).

In the face of recent climate change (e.g. Goodridge 1989), the study of the scale of vulnerability of human settlements is of great importance because it is one of the fundamental parameters within which to operate (Fagan 2009). This statement, linked to long-term phenomena, is now evident in the sudden climate changes recorded, as demonstrated by the recent deviations between weather conditions and those typical in many places—see for example the Legambiente report (Legambiente 2017) on Italian cities or (Giovannelli et al. 2012). Emblematic cases are now evident, from the influence on hunter populations in Greenland (e.g. Folger 2015), to the phenomena of desertification with the consequent repercussions on the habitability of large areas as highlighted in the World Atlas of Desertification (Charlet et al. 2015)—see also studies on Africa (e.g. Hamed et al. 2018; Saad et al. 2018) and large areas of Asia, such as the Mongolian Plateau (e.g. Wei et al. 2018).

An important in-depth analysis of the vulnerability threshold is linked to the conception that humanity has of environmental risk, as analysed in the sociological thought of Mary Douglas—see also (Jonas 1979). The author has defined *four myths* (Douglas and Wildavsky 1982), models of the same number of perceptions that populations show towards nature when subjected to perturbative phenomena, i.e. environmental risks. The first myth is represented by a *fatalist* vision. For fatalists, nature is capricious and, given its intrinsic unpredictability, its dynamics cannot be controlled. The second myth is of an *expansionist* vision. For this conception, nature is strong and capable of reacting to any disturbance by finding its initial equilibrium. There are no phenomena capable of changing its stability. The third myth is represented by a *communitarian* conception, which interprets nature as extremely fragile. Any deviation from the condition of equilibrium carries the risk of triggering irreversible degradation. Last but not least, the fourth myth is represented by a *managerial* vision. Managers perceive nature as strong within certain limits. In other words, nature is capable of coping with disturbances, but only within certain threshold values beyond which equilibrium is no longer possible. Referring to other texts for a detailed explanation of the individual myths (Chiesa 2010; Mela et al. 1998; Thompson 1983; Douglas and Wildavsky 1982), Table 1.1 aims to explore these visions, introducing the filter of climatic disturbances. It should be noted that the scheme analyses the perception of the variable and not the actual reactions of individuals to the climate parameter. In addition, it is possible to imagine the managerial myth with a representation in the form of a *bowl* (Chiesa 2010, 2017b) that adapts to a geometric representation of the forces and phenomena that interact on the ecosystem. These can be understood as vectors that share the centre of the bowl as one of the vertexes of direction, but present opposite sides or agreements with each other, depending on whether they represent movements away or approaches to the state of equilibrium, represented by the centre of the model. The scheme can be combined with a list of indicators of the ecological and environmental quality of

Table 1.1 The conceptions of nature in the face of disturbing phenomena—elaboration from (Chiesa 2010, 54) including climate issues and basing on (Douglas and Wildavsky 1982; Thompson 1983) "myths"

"Myths"	Implications				
	Social	Economics	Ecosystem	Technology	Climate
1	Fatalist		Out of my control, I don't care	Unlimited development	No implication, uncontrol-lable and evil nature
2	Expansionists	Unlimited growth in the invisible hand of the market	Space to colonise (species type I), Unlimited resources without predators	Unlimited development	There are no environmental and climate hazards, the threshold of vulnerability is infinite
3	Communitarists	Community dimension	Unstable balance, Migration or extinction in the case of a shift in balance	Local development sustainable and conscious approach	Minimum vulnerability threshold, any climate change has immediate repercussions
4	Directionists	Controlled growth	Climax, ecosystem of type III (Allenby and Cooper 1994)	Conscious development, biomimetic and sustainability approaches	Resistance to climate alterations within a certain threshold, beyond which there is the risk of collapse. Vulnerability threshold

an environment, taking care to place the value of equilibrium at the centre of the model. A parametric analysis can also be represented in the bowl diagram, working directly on the vectors or studying the consequences of a perturbative phenomenon. This model is suitable for virtually representing the vulnerability and resilience of settlements to climate change.

Moving away from the perceptual component of risk, it is stressed that the concept of the threshold of stability, which intervenes particularly in the case of the fourth myth, represents the domain in which climate does not constitute a vector of change in civilisation. However, while humanity has progressively increased its stability and resilience to the most recurrent disturbances, the current concentration and rate of anthropisation of the earth's soil has progressively reduced its ability to withstand climate change on a larger scale.

1.3 ICT Influences on Human Activities and Processes—A Four Axes Vision | It

The rapid development of ICT and their extensive presence in everyday life has given rise, as previously mentioned, to the start of a new phase of innovation and change in the relationship between man and technology (Braham and Hale 2007; Barker and Erickson 2005). However, it is necessary to investigate the effects of ICT on the redefinition of the conceptual, technical, philosophical and cultural tools and instruments of contemporary society. This study, which is mentioned in the report, referring to the research carried out in Europe by *The Onlife Initiative*, aims to avoid losing the possibility of managing machines and IT (Wiener 1966). The speed of change and spread of ICT has been so great as to highlight the need to completely redesign the conceptual tools for their understanding and conscious use (in terms of Re-designing and Re-engineering), linked to the need to address the digital revolution in the way we operate in scientific research as highlighted by (Floridi 2013; Nielsen 2012; Occelli and Staricco 2002).

The repercussions induced by innovations in ICT on human processes and activities can be localised around four macro-axes (Floridi 2013), as reported below and briefly described in the following points—see also the characterisation reported in (Chiesa 2016):

1. the reduction of the boundaries between the real and virtual worlds (Sakamoto and Ferré 2008; Sass and Oxman 2006; Oxman 2006; Mitchell 2005; Milgram and Colquhoun 1999; Negroponte 1995);
2. the hybridisation of the natural and artificial worlds (Hochberg et al. 2012; Gruber 2011; Chiesa 2010; Bar Cohen 2006; Benyus 1997; Gérardin 1968);
3. the transition from scarcity to abundance of information (Mayer-Schönberger and Cukier 2013; Nielsen 2012; Xu 2012; Weinberger 2012; Shirky 2010; Hashimoto and Dijkstra 2004; Wiener 1966);
4. the transition from the primacy of the entity to the primacy of interaction (City Form Lab; Sevtsuk and Amindarbari 2012; Weinberger 2012; Barabasi 2003).

1.3.1 Hybridisation Between Reality and Virtuality

The first axis is organised around the progressive overlap between the real and virtual worlds. We are witnessing a change in the relationship with computers, from "internal", characterised by the use of microprocessors and punch card machines, to "external", characterised by the use of personal computers (for a brief description of the history of computers see Gubitosa 2007). Contemporary technologies are, however, bringing us back to being "internal" through a process of *enveloping* reality due to sophisticated virtuality (Floridi 2013; The Onlife Initiative, e.g. see Floridi 2015). Additionally, the spread of technologies in the environment and more sophisticated virtual world modelling techniques is changing the way we design and interact with urban reality. Think of the concept of *augmented virtuality* (av), the world of video games (Chiesa 2014; Chiesa and La Riccia 2013) and that of *augmented reality* (ar), as in the mui project (Mobilier Urban Intelligent) promoted by the City of Paris and the projects of the cities of London, San Francisco, New York and Arles (see also Bimber and Raskar 2005). Studies of the spaces of integration between real and virtual worlds, in a perspective of *mixed reality* (mr), show how we see a progressive articulation of the three dimensions (av, ar and mr) in a context of taxonomic complexity. Based on studies conducted since 1994 by Milgran and other researchers, it is possible to describe the continuity between real and virtual environments as a continuous axis. Alongside this real-virtual axis runs a further continuous line representing the extension of knowledge of the world through its modelling, at the ends of which lie the unmodeled world (real) and the totally modelled world (virtual environment). Using a three-axis scheme it is also possible to highlight other typical characteristics of spaces along the real-virtual continuum, passing through the different *mixed realities* (axis 1), such as congruence (axis 2), understood as the type of control according to the visualisation method chosen, interacting between subjective view, global view and control on a personal or global scale, and *centricity* (axis 3), or the scope of vision, control and reference of the object. In the case of *centricity,* according to Milgram and Colquhoun (1999), a distinction is made between egocentric control, based on the point of view of the object, and exocentric control, based on a worldwide model of reference, for example when the camera used to move a robot uses satellite vision.

Alongside the modelling techniques in the digital age introduced earlier and the discourse about the tools of parametric virtualisation-design and algorithmic and architectural urban planning (Saleh and Al-Hagla 2012; Tang and Anderson 2011; Krish 2011), materialisation techniques based on a totally new way of interacting between bits and atoms are spreading, examples being the materialisation of the model with different scale control instruments (Zhang and Khoshnevis 2013; Anderson 2012; Lim et al. 2012; Khoshnevis 2004).

1.3.2 Hybridisation Between Nature and Artefact

The second axis contains hybrid innovations, which show how it is possible, with an awareness of scientific and technological knowledge, to create a new relationship between the natural and artificial worlds. This relationship can be interpreted in at least two ways: bionic hybridisation and biomimicry—see also Pagani et al. (2015). Bionic hybridisation (Gérardin 1968), similar to that which led Cathy Hutchinson in 2011 to control an external robotic arm with her own brain (Hochberg et al. 2012), the brain gate of which is an innovation that, according to Cardwell (1972), is part of the long history of artificiality. Biomimicry, on the other hand, consists in the emulation of the natural world which sees nature as the source of inspiration and tends towards a new phase in its relationship with it (Chiesa 2017b; Nagel 2014; Gruber 2011; DeLuca 2010; Pagani 2010; Chiesa 2010; Salvia et al. 2009; Bar Cohen 2006; Pagani 2008; Benyus 1997).

Alongside these two modes there is another axis of interaction between the artificial, human and natural worlds, based on the spread of specific ICT capable of interacting and integrate with the physical environment and the body in a perspective of enveloping related to the Internet of things, *kinect* technologies and some effects of augmented reality. The human being and the natural space become part and/or peripheral of computers that use physical space, of sensors capable of adhering to the body and of integrating into a system of sensors/actuators that includes man himself (Melgar and Diez 2012; Borensein 2012; Graham and Gowan 2009). Specific microprocessor cards have been developed to allow interaction with the clothes or the body, using sensing and actuating processes, such as continuous sound, visual and retroactive electrocardiograms. Moreover, some recent projects, such as Makey Makey—see for example https://makeymakey.com/ (viewed Dec. 2018)—show how it is possible to overcome the traditional constraint between computer peripherals, other real-world objects and human beings using a *Human interface device*. It is possible to use any object, connecting it with crocodile clips and wires, in place of a mouse, keyboard, joypad, piano keys and any other type of device you want to program. In these systems, the electrical circuit diagram also includes the human being, the user, also connected with a specific wire.

1.3.3 The Abundance of Data and Information—The Era of Big Data

The third axis is based on the growth, at all levels, of the amount of data produced annually, moving from a regime of scarcity to a situation of abundance. The current limits to the growth of data seem to be increasingly constituted by the limits imposed by thermodynamics and intelligence (Floridi 2013). In this context, the main problem related to big data is due to reduced storage capacity. Since 2009, more data have been produced than can be stored in HDD—just think of the increasing number pf Gb used annually with our Smartphone (e.g.: access to the Internet). The excess world (Anders 2010) becomes excess data, in a context of cognitive surplus (Shirky 2010) and network intelligence (Weinberger 2012) so that "the memory outperforms intelligence" (Floridi 2013). The repercussions on society due to the progressive expansion of big data and *cloud computing, collective intelligence and cloud manufacturing* techniques involve at least three changes in the way we analyse data and information, changing the way in which individuals relate to one another and organise themselves (Mayer-Schönberger and Cukier 2013):

- the logic of more. Big data allow the analysis of a growing number of datapoints, tending to overcome the distinction between sampling and total population and letting the granular structure leak out. "As of the 19th century, society has always relied on the use of samples when interpreting a large mass of numbers […] The use of all data allows us to see details that we could never have noticed when we were forced to examine limited quantities" (Mayer-Schönberger and Cukier 2013, 24);
- from the logic of accuracy, typical of the small data regime, to the logic of understanding the phenomenon at the macro level. "In the presence of fewer errors in sampling we can accept more errors in measurement […] Exactness requires the ultra-analytical processing of data […] In the presence of big data, we are satisfied with a general trend […] but we do not completely forego precision: we only forego the devotion we have always felt towards it" (Mayer-Schönberger and Cukier 2013, 24–25). In other words, although "a given reading might be wrong […] the aggregate of lots of readings will provide a more comprehensive picture [taking into account the fact that] big data turn figures into something probabilistic rather than precise" (Mayer-Schönberger and Cukier 2013, 54–55). This is why, in cases where precision is required, it is important to use big data as a way to get closer to reality than traditional means (Nielsen 2012), although this is mainly valid for the launch of spacecraft;
- from the logic of why to the logic of what. "In the world of big data […] we will no longer have to focus on randomness; instead, the data will show us trends and correlations that offer us original and precious indications. The correlations will not tell us exactly why something happens but will at least warn us that [in fact] the correlations are useless in a world dominated by small data, but in the context of big data they emerge in all their importance" (Mayer-Schönberger and Cukier 2013, 26, 76). As data points increase, the risk of false correlations also increases. However, correlations exist and can be demonstrated, unlike some random links. From the research point of view, correlations can be applied in the identification of the most important variables to be analysed through experiments aimed at identifying the cause. In the field of energy simulation, too, we are beginning to see the potential to develop energy performance benchmarking methods from clustering algorithms based on a large number of potential indicators (e.g. Gao and Malkawi 2014). Simple applications in low-energy design can concern the optimisation of specific design choices—e.g. U-value or the percentage of window opening, or even the definition of ac/h (air changes per hour) for optimising the use of natural controlled ventilation—by using massive dynamic energy simulations performed through scripting codes (e.g. Chiesa et al. 2018, 2019; Grosso et al. 2019).

It is fundamental, however, to remember that "big data have a solid theoretical foundation" (Mayer-Schönberger and Cukier 2013, 101) and will not contribute to putting an end to the theories. On the one hand, they require a high level of mathematical, statistical and IT skills and knowledge, and on the other, they require strong theoretical foundations because the methods used have a strong influence on the results. "Conditioning starts with the way we select data […] our choices are conditioned by theories" (Mayer-Schönberger and Cukier 2013, 101). However, our way of understanding reality and science is radically changing (Nielsen 2012), in a landscape that could lead to a new modernity or non-modernity (Latour 2008, 2009) of society characterised by the progressive coming together of scientists and the general public, with regard to randomness, but by a deep estrangement with regard to the definition of the thing.

The three changes introduced are directly related to the *long tail* concept introduced by Anderson. This concept, already applied to many areas of knowledge by the author himself, is now directly connected with the economy, with the materialisation and production of goods and objects, in a new vision between the real world and the virtual world, which has significant similarities with what is described with regard to the second digital age and the concept of information (Anderson 2009, 2010, 2012). The increase in data, and in the number and frequency of datapoints, makes it possible to define areas in which the cost of specialist knowledge is decreasing as the greater amount of information and knowledge is moving towards the long tail (Weinberger 2012). However, as Nielsen reminds us in *The new ways of scientific discovery,* their use must, at least in part, refer to new figures of data writers capable of avoiding delegation to a machine or software the decisions "of our conduct without first having studied the laws that govern its behaviour, and without knowing with certainty that this behaviour will be based on principles that we can accept! Like the genie in the bottle, the machine that can learn and make decisions on the basis of that acquired knowledge will not in any way be obliged to decide in the same way that we would have decided ourselves, or at least in a way acceptable to us" (Wiener 1966, 228). In this sense, process management, objectives and methodology cannot be delegated.

1.3.4 From the Primacy of Entities to the Primacy of Interactions

The fourth axis deals with the issue of the change of centrality that has occurred between the nodes and networks. While, historically, networks have been organised around nodes, tessellating space, today, the node is born where two links meet (*City Form Lab*). The physical and virtual network becomes the space for data and object management, passing from an *object-based* concept to an *interaction-based* vision (*The Onlife Initiative*).

The increase in virtualisation technologies, and in data mining, data production and data storage techniques, changes the way we think and interpret urban processes in digital and smart cities (Sevtsuk and Amindarbari 2012).

The interaction connects with the new technologies of materialisation of the first axis in a perspective of *cloud manufacturing* (Fisher et al. 2018; Xu 2012; Anderson 2012), but also allows research based on *citizen science* such as *Galaxy Zoo* for example, to play a game of world chess against Kasparov and to develop open source software. In other words, as in the case of chess "Irina Krush was inferior to Kasparov in almost all areas of chess, but in that particular micro-skill she exceeded the great champion [precisely because interaction means allowing] the discovery of these latent micro-skills, to use them when necessary" (Nielsen 2012, 31). Another example of interaction is the *Network Challenge* organised by DARPA on the anniversary of the creation of ARPANet, as we are reminded by Weinberger (2012). The contest consisted in being the first to find 10 large red balls positioned in different locations in the USA. The aim of the initiative was to test the ability of online platforms such as social networks to collect and share information. The competition was won by a MIT group in just nine hours, thanks to a website that allowed the assignment of a percentage of the prize to those who indicated the location of the balls or helped get information (Weinberger 2012, 73). The project promoted by DARPA showed that the power of interaction in a network system makes it possible to quickly solve a problem that would be impossible for a single entity to solve. This network articulation, which works on different scales and is proving to be cumulative, has profound repercussions on expertise, which is beginning to be articulated in connective repercussions (Weinberger 2012; Anderson 2010).

References

Adriaenssens S, Gramazio F, Kohler M, Mengers A, Pauly M (eds) (2016) Advances in architectural geometry 2016. vdf Hochschulverlag AG an der ETH Zurich, Zurich

Allenby BR, Cooper WE (1994) Understanding industrial ecology from a biological systems perspective. Total Qual Environ Manage 3(3):343–354

Alvioli M, Melillo M, Guzzetti F, Rossi M, Palazzi E, von Hardenberg J, Brunetti MT, Perucciani S (2018) Implications of climate change on landside hazard in Central Italy. Sci Total Environ 630:1528–1543

Anders G (2010) Eccesso di mondo. Processi di globalizzazione e crisi del sociale. Millepiani 16. Mimesis, Milano, pp 7–28

Anderson C (2012) Makers: the New Industrial Revolution. Random House Business Books, London

Anderson C (2010) La coda lunga. Da un mercato di massa a una massa di mercati. Codice edizioni, Torino [or (ed) (2006) The long tail; why the future of business is selling less of more. Hyperion, New York]

Anderson C (2009) Gratis. RCS libri, Milano [or (ed) (2009) Free: the future of a radical price. Hyperion, New York]

Anderson C (2006) The long tail; why the future of business is selling less of more. Hyperion, New York

Aqrawi AAM (2001) Stratigraphic signatures of climatic change during the Holocene evolution of the Tigris-Euphrates delta, lower Mesopotamia. Global Planet Change 28:267–283

Bagg AM (2000) Irrigation in Northern Mesopotamia: water for the Assyrian capitals (12th–7th centuries BC). Irrigat Drain Syst 14:301–324

Ballesteros M et al (eds) (2008) Verb Crisis, Architecture Boogazine. Actar, Barcelona

Barabasi AL (2003) Link. La scienza delle reti. Einaudi, Torino [or (ed) (2002) Linked: the new science of networks. Perseus Books Group, New York]

Bar Cohen Y (ed) (2006) Biomimetics: biologically inspired technologies. CRT Taylor & Francis Group, Boca Raton

Barker JA, Erickson S (2005) Five regions of the future: preparing your business for tomorrow's technology revolution. Penguin Group, New York

Beckers B (ed) (2012) Solar energy at urban scale. Wiley-ISTE, New York

Benyus JM (1997) Biomimicry: innovation inspired by nature. HarperCollins, New York

Block P, Knippers J, Mitra NJ, Wang W (eds) (2014) Advances in architectural geometry 2014. Springer, Cham

Bimber O, Raskar R (2005) Spatial Augmented reality: merging real and virtual worlds. AK Peters, Wellesley

Bocco A, Cavaglià G (2008) Cultura tecnologica dell'architettura. Pensieri e parole, prima dei disegni. Carocci, Roma

Borensein G (2012) Making things see. O'Reilly, Sebastopol

Bottaccioli L, Estebsari A, Patti E, Pons E, Acquaviva A (2017) A novel integrated real-time simulation platform for assessing photovoltaic penetration impacts in smart grids. Energy Procedia 111:780–789

Bottaccioli L, Patti E, Macii E, Acquaviva A (2018) Distributed Infrastructure for multi-energy-systems modelling and co-simulationin urban districts. In: Proceedings of the 7th conference on smart cities and green ICT systems—SMARTGREENS 2018, Funchal, Madeira, Portugal, 16–18 Mar, pp 262–269

Braham WW, Hale JA (eds) (2007) Rethinking technology. A reader in architectural theory. Routledge, London and New York

Broecker W (1997) Future directions of paleoclimate research. Quatern Sci Rev 16:821–825

Butera FM (2004) Dalla caverna alla casa ecologica. Storia del comfort e dell'energia, Edizioni Ambiente, Milano

Cardwell DSL (1972) Technology, science and history. Heinemann, London

Carrara G, Fioravanti A, Loffreda G, Trento A (2014) Conoscere collaborare progettare. Gangemi, Roma

Cavaglià G, Ceragioli G, Foto M, Maggi PN, Matteoli L, Ossola F (1975) Industrializzazione per programmi. Strumenti e procedure per la definizione dei sistemi di edilizia abitativa. Studi e Ricerche RDB, Piacenza

Ceccato C et al (eds) (2010) Advances in architectural geometry 2010. Springer, Wien

Celento D (2007) Innovate or Perish. New technologies and architecture's futures. Harvard Design Magazine 27:1–9

Chamberlin S (2009) The transition timeline, for a local, resilient future. Green Books, Foxhole

Charlet M, Reynolds J, Hutchinson C, Hill J, von Maltitz G, Sommer S, Ajai, Fensholt R, Horion S, Shepherd G, Weynants M, Kutnjak H, Smid M (2015) World atlas of desertification. Mapping land degradation and sustainable land management opportunities, 3rd edn. European Commission, Joint Research Centre, Luxemburg

Chiesa G (2019) Calculating the geo-climatic potential of different low-energy cooling techniques. Build Simul 12(2):157–168

Chiesa G (2017a) Explicit-digital design practice and possible areas of implication. Techne 13:236–242

Chiesa G (2017b) Biomimetics. Technology and innovation for architecture. Celid, Torino

Chiesa G (2016) Model, digital technologies and datization. Toward an explicit design practice. In: Pagani R, Chiesa G (eds) Urban data. Tools and methods towards the algorithmic city. FrancoAngeli, Milano, pp 48–81

Chiesa G (2014) Luoghi e non luoghi di ritrovo giovanile. Insegnare 39, 9 pp [ed (or) 2012]

Chiesa G (2013) Il ritorno del dato. Dati città piattaforme tra reale e virtuale. In: Proceedings OSDOTTA 2013, Firenze University Press, Firenze [under publication]

Chiesa G (2010) Biomimetica, tecnologia e innovazione per l'architettura. Celid, Torino

Chiesa G, Acquaviva A, Grosso M, Bottaccioli L, Floridia M, Pristeri E, Sanna EM (2019) Parametric optimization of window-to-wall ratio for passive buildings adopting a scripting methodology to dynamic-energy simulation. Sustainability 11:30. https://doi.org/10.3390/su11113078

Chiesa G, Casetta E (2012) Parametric CAD modelling and scripting for architectural technology [unpublished paper]

Chiesa G, De Paoli O (2014) Modelling for project design: instruments for a sustainable and integrated design. SMC—Sustain Mediterr Constr 1:115–119

Chiesa G, Di Gioia A (2011) Rappresentare il territorio della contemporaneità: la fotografia ambientale come supporto all'analisi territoriale. Planum, 11 pp
Chiesa G, Grosso M (2019) Meta-design approach to environmental building programming for passive cooling of buildings. In: Sayigh A (ed) Sustainable building for a cleaner environment. Springer, Cham, pp 285–296
Chiesa G, Grosso M (2017) An environmental and technological approach to architectural programming for school facilities. In: Sayigh A (ed) Mediterranean green buildings and renewable energy. Springer, Amsterdam, pp 701–715
Chiesa G, Grosso M, Acquaviva A, Makhlouf B, Tumiatti A (2018) Insulation, building mass and airflows—provisional and multi-variable analysis. SMC-Sustain Mediterr Constr 8:36–40
Chiesa G, La Riccia L (2013) Dalla rappresentazione alle rappresentazioni di paesaggi e territori. Planum 27(2)
Clark S (2014) Artificial intelligence could spell end of human race—Stephen Hawking, The Guardian, 2 Dec. Accessible at: https://www.theguardian.com/science/2014/dec/02/stephen-hawking-intel-communication-system-astrophysicist-software-predictive-text-type, last view June 2019
Cohen G, Mijkamp P (2002) Information and communication technology policy in European cities: a comparative approach. Environ Plan B: Plan Des 29:729–755
Coppens Y (2006) Storia dell'uomo e cambi di clima. Jaca Book, Milano [or (ed) (2006) Histoire de l'homme et changements climatiques. Collège de France et Fayard, Paris]
Crutzen PJ (2005) Benvenuti nell'antropocene. Mondadori, Milano
Crutzen PJ, Stoermer EF (2000) The "Anthropocene". IGBP Newsletter 41:17–18
DeLuca D (2010) Biomimicry: innovation inspired by nature. In: Matteoli L, Pagani R (eds) CityFutures. Architecture design technology for the future of the cities. Hoepli, Milano, pp 275–280
Dollens D (2006) The Pangolin's guide to biomimetics and digital architecture. Sites Books, Santa Fe
Douglas M, Wildavsky A (1982) Risk and culture. An essay on the selection of technological and environmental danger. University of California Press, Berkeley
Droege P (2006) The renewable city: A comprehensive guide to an urban revolution. Wiley Academy, Milton, Queensland
Eilouti B (2018) Concept evolution in architectural practice: an octonary framework. Front Archit Res 7:180–196
Erell E, Pearlmutter D, Williamson T (2015) Urban microclimate: designing the spaces between buildings. Routledge, London
Fagan B (2009) La lunga estate. Come le dinamiche climatiche hanno influenza la civilizzazione. LeScienze, Roma [or (ed) (2004) The long summer: how climate changed civilization. Basic Books, New York]
Fagan B (2002) The little ice age: how climate made history 1300–1850. Basic Books, New York
Ferré et al (eds) (2005) Verb conditioning, architecture boogazine. Actar, Barcelona
Ferré et al (eds) (2004) Verb connection, architecture boogazine. Actar, Barcelona
Ferré et al (eds) (2003) Verb matters, architecture boogazine. Actar, Barcelona
Fisher O, Watson N, Porcu L, Bacon D, Rigley M, Gomes RL (2018) Cloud manufacturing as a sustainable process manufacturing route. J Manuf Syst 47:53–68
Floridi L (2017) La quarta rivoluzione industriale. Come l'infosfera sta trasformando il mondo. Raffaello Cortina, Milano
Floridi L (ed) (2015) The onlife Manifesto. Being Human in a hyperconnected era. Springer, Cham
Floridi L (2013) The on life manifesto. Seminary, Centro Nexta about Internet and Società. Politecnico di Torino, DAUIN, Turin, 24 May 2013, Italy
Foley SF et al (2013) The palaeoanthropocene—the beginnings of anthropogenic environmental change. Anthropocene 3:83–88
Folger T (2015) Melting away: for Greenland's hunters, fading sea ice could mean moving beyond a traditional way of living. In: National geographic, pp 98–119

Gao X, Malkawi A (2014) A new methodology for building energy performance benchmarking: an approach based on intelligent clustering algorithm. Energy Build 84:607–616
Gérardin L (1968) La Bionique. Fayard, Paris
Geymonat L (1986) Storicità e attualità della cultura scientifica. Insegnare 11:16
Giovannelli G et al (2012) Innovazione e sostenibilità nel settore edilizio; "costruire il futuro", Primo Rapporto dell'Osservatorio congiunto Fillea CGIL—Legambiente. Roma, last view Apr 2019. http://www.itaca.org/documenti/news/RAPPORTO%20FILLEA-LEGAMBIENTE.pdf
Goodridge J (1989) Air temperature trends in California, 1916 to 1987. Chico, CA, p 95928
Graham B, Mc Gowan K (2009) Mind performance projects for the evil genius. Tab Books, Blue Ridge Summit
Grosso M (2005) Valutazione dei caratteri energetici ambientali nel metaprogetto. In: Grosso M, Peretti G, Piardi S, Scudo G (eds) Progettazione ecocompatibile dell'architettura. Esselibri, Napoli, pp 307–336
Grosso M (1998) Urban Form and renewable energy potential. Renew Energy 15:331–336
Grosso G, Acquaviva A, Chiesa G, da Fonseca H, Bibak Sareshkek SS, Padilla MJ (2019) Ventilative cooling effectiveness in office buildings: a parametrical simulation. In: Proceedings of the 39th AIVC—7th TightVent & 5th venticool conference—smart ventilation for buildings, Antibes Juan-Les-Pins Conference Centre, France, 18–19 Sept 2018, pp 780–788. ISBN: 2-930471-53-2. Available at https://www.aivc.org/download/aivc2018-proceedings.pdf, last view May 2019
Gruber P (2011) Biomimetics in architecture. Springer, Wien-New York
Gubitosa C (2007) Hacker, Scienziati e Pionieri. Stampa Alternativa/Quali Equilibri, Viterbo
Haarsma RJ, Roberts MJ, Vidale PL, Senior CA, Bellucci A, Bao Q, Chang P, Corti S, Fuˇckar NS, Guemas V, von Hardenberg J, Hazeleger W, Kodama C, Koenigk T, Leung LR, Lu J, Luo J-J, Mao J, Mizielinski MS, Mizuta R, Nobre P, Satoh M, Scoccimarro E, Semmler T, Small J, von Storch J-S (2016) High resolution model intercomparison project (HighResMIP v1.0) for CMIP6. Geosci Model Dev 9:4185–4208
Hamed Y, Hadji R, Redhaounia B, Zighmi K, Baali F, El Gayar A (2018) Climate impact on surface and groundwater in North Africa: a global synthesis of findings and recommendations. Euro-Mediterr J Environ Integr 3:15
Hashimoto S, Dijkstra R (2004) Chip city. In: Ferré et al. (eds) Verb connection, architecture boogazine. Actar, Barcelona, pp. 46–53
Hawkes D, Owers J, Rickaby P, Steadman P (eds) (1987) Energy and built form. Butterworths, London
Hesselgren L, Sharma S, Wallner J, Baldassini N, Bompas P, Raynaud J (eds) (2012) Advances in architectural geometries 2014. Springer, Wien
Hochberg et al (2012) Reach and grasp by people with tetraplegia using a neurally controlled robotic arm. Nature 485:372–375
Holloway A et al (2012) Silkworm. Manual, last view Jan 2014. http://projectsilkworm.com/download
Hopkins R (2008) The transition handbook. From oil dependency to local resilience. Green Books, Foxhole
How D, Stouffs R (2018) An algorithmic design grammar for problem solving. Autom Constr 94:417–437
Hwang I et al (eds) (2006) Verb nature, architecture boogazine. Actar, Barcelona
Jonas H (1979) Das Prinzip Verantwortung: Versuch einer Ethik für die technologische Zivilisation. Frankfurt [ITA translation (1990) Il principio di responsabilità. Einaudi, Torino]
Khosnevis B (2004) Automated construction by contour rafting—related robotics and information technologies. Autom Constr 13:5–19
Kocaturk T (2012) Reading/understanding the digital revolution in architecture. Lecture at the excellence Ph.D. course in Real Design, virtual prototyping, digital interaction. Politecnico di Torino, Turin, Italy
Krish S (2011) A practical generative design method. Comput Aided Des 43:88–100

Kurzweil R (2008) La singolarità è vicina. Apogeo, Milano [or (ed) (2005) The singularity is near: when humans trascend biology, Viking, New York]
Latour B (2008) Disinventare la modernità. Conversazioni con François Ewald. Elèuthera, Milano [or (ed) (2005) Un monde pluriel mais commun. Entretien avec François Ewald. Editions de L'aube, La Tour-d'Aigues]
Latour B (2009) Non siamo mai stati moderni. Elèuthera, Milano [or (ed) (1991) Nous n'avons jamais été modernes. Essai d'anthropologie symétrique. La Découverte, Paris]
Corbusier L (1923) Vers une architecture. G. Crès, Paris
Lee et al (2002) Developing a vision for a nD modelling tool. In: Agger C et al (eds) Distributing knowledge in buidling. CIB W78 conference proceedings, vol 2. Aarhus School of Architecture, Denmark
Legambiente (2017) Le città alla sfida del clima. Gli impatti dei cambiamenti climatici e le politiche di adattamento. Legambiente, Roma
Lim S et al (2012) Developments in construction-scale additive manufacturing processes. Autom Constr 21:262–268
Lynch-Stieglitz J, Curry WB, Slowey N (1999) Weaker Gulf Stream in the florida straits during the last glacial maximum. Nature 402:644–648
Lynn G (1999) Animate form. Princeton Architectural Press, New York
Magnaghi A (ed) (1994) Il territorio dell'abitare. Lo sviluppo locale come alternativa strategica. FrancoAngeli, Milano
Maldonado T (2007) Reale e virtuale, 2nd edn. Feltrinelli, Milano
Mateos JL (ed) (2007) Natural metaphor. An anthology of essays on architecture and nature. Actar, Barcelona—New York
Mayer-Schönberger V, Cukier K (2013) Big Data. Una rivoluzione che trasformerà il nostro modo di vivere e già minaccia la nostra libertà. Garzanti, Milano [ed (or) (2013) Big data: a revolution that will transform how we live, work, and think. Houghton Mifflin Harcourt, Boston]
Mela A, Belloni MC, Davico L (1998) Sociologia dell'ambiente. Carocci, Roma
Melgar ER, Diez CC (2012) Arduino and Kinect Projects. Design, Build, Blow Their Minds. Apress, Springer Science Distribution, New York
Milgram P, Colquhoun H (1999) Chapter 1: a taxonomy of real and virtual world display integration. In: Ohta Y, Tamura H (eds) Mixed reality: merging real and virtual worlds. Springer, New York, pp 5–30
Mitchell M (1998) An introduction to genetic algorithms. MIT Press, Cambridge
Mitchell TM (1997) Machine learning. McGraw-Hill, New York
Mitchell WJ (2005) Construction complexity. In: Martens B, Brown A (eds) Computer aided architectural design futures 2005. Springer, Netherlands, pp 41–50
Monod J (1970) Il caso e la necessità. Saggio sulla filosofia naturale della biologia contemporanea. Arnoldo Mondadori Editore, Milano [or (ed) (1970) Le Hasard et la Nécessité: Essai sur la philosophie naturelle de la biologie moderne. Éditions du Seuil, Paris]
Moore AMT (2000) Village on the Euphrates: from Foraging to farming at Abu Hureyra. Oxford University Press, New York
Nagel JKS (2014) A thesaurus for bioinspired engineering design. In: Goel A, McAdams DA, Stone RB (eds) Biologically inspired design. Computational methods and tools. Spinger, London
Nakano A, Bueno B, Norford LK, Reinhart C (2015) Urban weather generator—a novel workflow for integrating urban heat island effect within urban design process. Building Simulation 2015. http://urbanmicroclimate.scripts.mit.edu/publications.php, last view Dec 2018
Nakano A (2015) Urban weather generator user interface development: towards a usable tool for integrating urban heat island effect within urban design process. MS Thesis, MIT Building Technology
Negroponte N (1995) Essere digitali. Sperling & Kupfer Editori, Milano [or (ed) (1995) Being digital, Alfred A. Knopf, New York]

Nielsen M (2012) Le nuove vie della scoperta scientifica. Come l'intelligenza collettiva sta cambiando la scienza. Einaudi, Torino [or (ed) (2012) Reinventing discovery: the new era of networked science. Princeton University Press, Princeton]
Occelli S, Staricco L (2002) Nuove tecnologie di informazione e di comunicazione e la città. FrancoAngeli, Milano
Osello A, Acquaviva A, Del Giudice M, Patti E, Rapetti N (2016) District information models. The DIMMER project: BIM tools for the urban scale. In: Pagani R, Chiesa G (eds) Urban data. Tools and methods towards the algorithmic city. FrancoAngeli, Milano, pp 231–261
Oxman R (2006) Theory and design in the first digital age. Des Stud 27:229–265
Pagani R (2010) Architettura e sostenibilità: utopia o nuovo impegno di progetto?. In: Chiesa G (ed) Biomimetica, tecnologia e innovazione per l'architettura. Celid, Torino, pp 9–14
Pagani R (2008) Architettura e sostenibilità: utopia o un nuovo impegno di progetto?. Prolusion presented at the inauguration of the Academic Year 2007–2008, Politecnico di Torino, Turin, Italy, last view May 2013. http://www.polito.it/ateneo/storia/inaugurazioni/2008/intervento_pagani.pdf
Pagani R, Chiesa G (eds) (2016) Urban data. Tools and methods towards the algorithmic city. FrancoAngeli, Milano
Pagani R, Chiesa G, Tulliani J-M (2015) Biomimetica e Architettura. Come la natura domina la tecnologia. FrancoAngeli, Milano
Palazzi E, von Hardenberg J, Provenzale A (2013) Precipitation in the Hindu-Kush Karakoram Himalaya: observations and future scenarios. J Geophys Res Atmos 118:85–100
Palazzi E, Filippi L, von Hardenberg J (2017) Insights into elevation-dependent warming in the Tibetan Plateau-Himalayas from CMIP5 model simulations. Clim Dyn 48:3991–4008
Pancini A et al (2010) Progettazione generativa, last view Dec 2011 [no longer available]. http://issuu.com/chupa.group/docs/progettazione_generativa
Penttilä H (2006) Describing the changes in architectural information technology to understand design complexity and free-form architectural expression. ITcon 11 (Special Issue the effects of CAD on Building Form and Design Quality), pp 395–408
Rhomer E (1964) Vers l'unité du monde. Les métamorphoses du paysage: l'ère industrielle. Documentary film of the series En profil dans le texte. Duration 22:23
Robinson D (2011) Computer modelling for sustainable urban design. Routledge, London
Rothenberg J (1989) The nature of modelling. In: Widman LE, Loparo KA, Nielson NR (eds) Artificial intelligence, simulation & modelling. Wiley, New York, pp 75–92
Ruiz-Montiel M, Belmonte M-V, Boned J, Mandow L, Millan E, Badillo AR, Perez-de-la-Cruz J-L (2014) Layered shape grammar. Comput Aided Des 56:104–119
Saad SAM, Seedahmed AMA, Ahmed A, Ossman SAM, Eldoma AMA (2018) Combating desertification in Sudan: experiences and lessons learned. In: Proceedings of the international conference public private partnerships for the implementation of the 2030 Agenda for Sustainable Development, Geneva, WASD, pp 141–155
Sakamoto T, Ferré A (eds) (2008) From control to design. Parametric/algorithmic architecture. Actar, Barcelona
Salazar J et al (eds) (2001) Verb processing, architecture boogazine. Actar, Barcelona
Saleh MM, Al-Hagla KS (2012) Parametric urban comfort envelope an approach toward a responsive sustainable urban morphology. Eng Technol 71:563–570
Salvia G, Rognoli V, Levi M (2009) Il progetto della natura: gli strumenti della biomimesi per il design. FrancoAngeli, Milano
Sass L, Oxman R (2006) Materializing design: the implications of rapid prototyping in digital design. Des Stud 27:325–355
Sassen S (2018) Cities in a world economy, 5th edn. Sage Publications, Newbury Park, California
Sassen S (2010) Le città nell'economia globale, 3rd edn. Il Mulino, Bologna [or (ed) (1994) Cities in a world economy. Pine Forge Press, Sage, Thousand Oaks]
Sevtsuk A, Amindarbari R (2012) Measuring growth and change in East-Asian cities. Progress report on urban form and land use measures. The World Bank, City Form Lab, Singapore

Shirky C (2010) Surplus Cognitivo. Creatività e generosità nell'era digitale. Codice Edizioni, Torino [or (ed) (2010) Cognitive surplus: creativity and generosity in a connected age. Penguin Group, London]
Slaughter RA (2012) Welcome to the Anthropocene. Futures 44:119–126
Sophocles (about 442 BC) Antigone, Chorus
Sutherland IE (2003) Sketchpad: a man-machine graphical communication system, Technical Report No. 574, UCAM-CL-TR-574, University of Cambridge, Cambridge. Accessible at: https://www.cl.cam.ac.uk/techreports/UCAM-CL-TR-574.pdf, last view June 2019
Sutherland IE (1963) Sketchpad, a man-machine graphical communication system, Ph.D. Thesis, Supervisor Shannon CE, MIT, Boston. Accessible at: http://images.designworldonline.com.s3.amazonaws.com/CADhistory/Sketchpad_A_Man-Machine_Graphical_Communication_System_Jan63.pdf, last view June 2019
Syvitski J (2012) Anthropocene: an epoch of our making. Global Change 78:12–15
Tang M, Anderson J (2011) Information urbanism. Parametric urbanism in junction with GIS data processing & fabrication. In: Proceedings of ARCC 2011—considering research: reflecting upon current themes in Architectural Research, Detroit, MI, 20–24 Apr 2011
Terzidis K (2006) Algorithmic architecture. Elsevier, Architectural Press, Oxford
Thompson M (1983) A cultural basic for comparison. In: Kunreuther H, Linnerooth J (eds) Risk analysis and decision process: the siting of liquefied energy gas facilities in four countries. Springer, Berlin
Todd NJ, Todd J (1993) From eco-cities to living machines. Principles of ecological design. North Atlantic Books, Berkeley
Turri E (1998) Il paesaggio come teatro. Dal territorio vissuto al territorio rappresentato, Marsilio, Venezia
Ur JA (2010) Cycles of civilization in Northern Mesopotamia, 4400–200 BC. J Archaeol Res 18:387–431
van Haren R, Haarsma RJ, de Vries H, van Oldenborgh GJ, Hazeleger W (2015) Resolution dependence of circulation forced future central European summer drying. Environ Res Lett 10(5):1–8
Vitta M (2008) Dell'abitare. Corpi, spazi, oggetti, immagini. Einaudi Editore, Torino
Wei H, Wang J, Cheng K, Li G, Ochir A, Davaasuren D, Chonokhuu S (2018) Desertification information extraction based on feature space combinations on the Mongolian Plateau. Remote Sens 10:17
Weinberger D (2012) La stanza intelligente. La conoscenza come proprietà della rete. Codice edizioni, Torino [or (ed) (2011) Too big to know: rethinking knowledge now that the facts aren't the facts, experts are everywhere, and the smartness person in the room is the room. Basic Book, New York]
Weiss H et al (1993) The genesis and collapse of third Millennium North Mesopotamian civilization. Science 261:995–1004
Wiener N (1966) Introduzione alla cibernetica. L'uso umano degli esseri umani. Editore Boringhieri, Torino [or (ed) (1950) The human use of human beings: cybernetics and society. Houghton Mifflin Company, Boston]
Wiener N (1950) The human use of human being. Houghton Mifflin, Boston
Woodbury R (2010) Elements of parametric design. Routledge, London and New York
Xu X (2012) From cloud computing to cloud manufacturing. Robot Comput Integr Manuf 28:75–86
Yang PP-J (2015) Energy resilient urban form: a design perspective. Energy Procedia 75:2922–2927
Zhang J, Khoshnevis B (2013) Optimal machine operation planning for construction by Contour Crafting. Autom Constr 29:50–67
https://makeymakey.com/, last view Dec 2018
http://senseable.mit.edu/, last view Apr 2019
http://cityform.mit.edu/, last view Apr 2019

Chapter 2
Modelling Reality, Modelling Virtuality

Open source lighting (LED) and shading (screen) control system.

2.1 Introduction

The progressive increase in relations between the virtual world and the real world, in conjunction with the spread of new ICT, is radically changing the ways the two worlds interact, giving rise to hybrid realities, as documented in numerous studies conducted in recent years in literature and in multiple applications presented in architecture. An interesting example is the possibility of controlling the real world through parametric digital representations that act on specific actuators and, vice versa, the possibility of influencing a virtual model starting from what happens in the real world. To this end, an example of a sensor-actuator system based on the *Arduino* open source and open hardware platform has been developed for the control of lighting within a simulated environment. We also studied the connection between the real home automation system and its virtual double, represented in *Rhinoceros 5*, a CAD environment capable of real time updates. The application presented uses the brightness values monitored to simultaneously manage a LED artificial lighting system and a mobile screen operated by a servomotor. This is an example of a use of monitored data for the control of a system. The open source nature of the hardware developed anticipates the contents of Chap. 3 on the production of real time data, foreseeing possible interactions between local home automation systems and the urban context, such as the presence of pollutants and the reduction of solar radiation in relation to the temperature of a block of buildings. The relationship with that developed in Chap. 3 on real time data extends further to the connection between the real and virtual worlds between the physical demonstrator and its virtual model.

G. Chiesa, *Technological Paradigms and Digital Eras*,
PoliTO Springer Series, https://doi.org/10.1007/978-3-030-26199-3_2

2.1.1 *Objectives*

The chapter has two general goals:

- the development and study of a simple and replicable home automation system for the control of illumination within a model programmed using an open source software and hardware solution. This system maximises the natural light component while minimising free contributions when the overall solar radiation is greater than that necessary to guarantee visual comfort;
- the implementation of a bridge between models, belonging to the real world, and a digital CAD model of the same environment, updated in real time.

The microcontroller board performs digital management between real world, sensors, data, analysis, information, optimisation, actuator and real world. In addition, the platform manages a serial port and some pins to send information to a script developed in a CAD environment. By adequately modifying the software loaded in the board, the information flow can be adapted, allowing the real model to adapt to the virtual one.

The example given in the chapter aims to make a simple home automation application understandable and replicable, however, if we wanted to expand the scope of these applications towards the development of a complete IoT (Internet of Things) system, a more detailed design of the methods of connection of the different elements involved would be more appropriate. Furthermore, in this example the internet connection, the definition of slave and master server units or the platform with APIs and library database to recognise and interact with devices are not defined.

An example of an IoT system is shown in Fig. 2.1. This system is based on a server unit composed by a Raspberry Pi B+ that interacts with various sensors/actuators that were previously described in the library definition phase.

2.2 Description and Development of the Demonstrator and Related Control Systems

The case study consists of a wooden demonstrator on which the applicability of a control system to sensors and actuators based on open source development platforms has been tested. Specifically, the demonstrator, consisting of a rectangle measuring $20 \times 21.20 \times 21.20$ cm, has a single full-height opening with a width of 8.10 cm. The mobile screen consists of a single vertical slat, designed to block the entire opening and connected to a motor for its movement. The components that make up the system hardware are shown in Fig. 2.2.

A detailed description of the hardware components and software development required to operate the demonstrator is shown below.

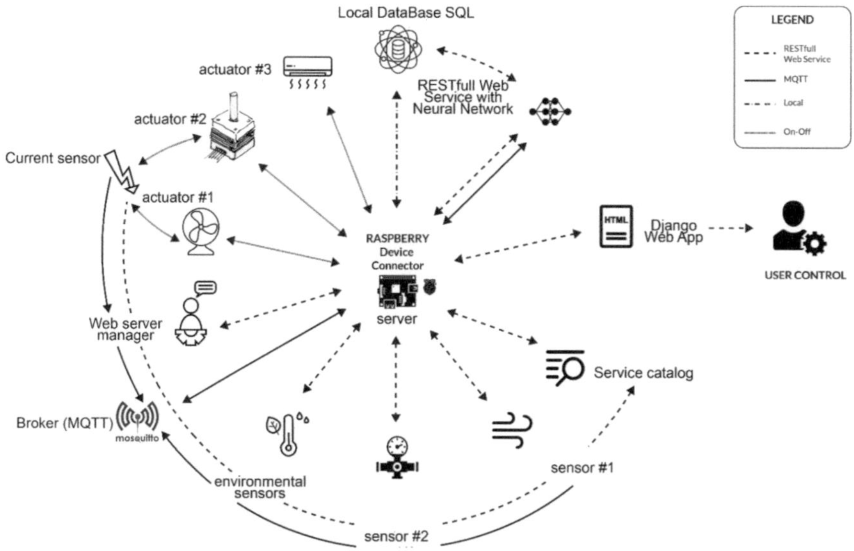

Fig. 2.1 Sample scheme of an IoT system—Interdisciplinary Course Project, Politecnico di Torino, A.Y. 2018–19—students: Gianluca Perna, Annalisa Parisi, Gianmarco Dragonetti; proff. Chiesa, Grosso, Dovis

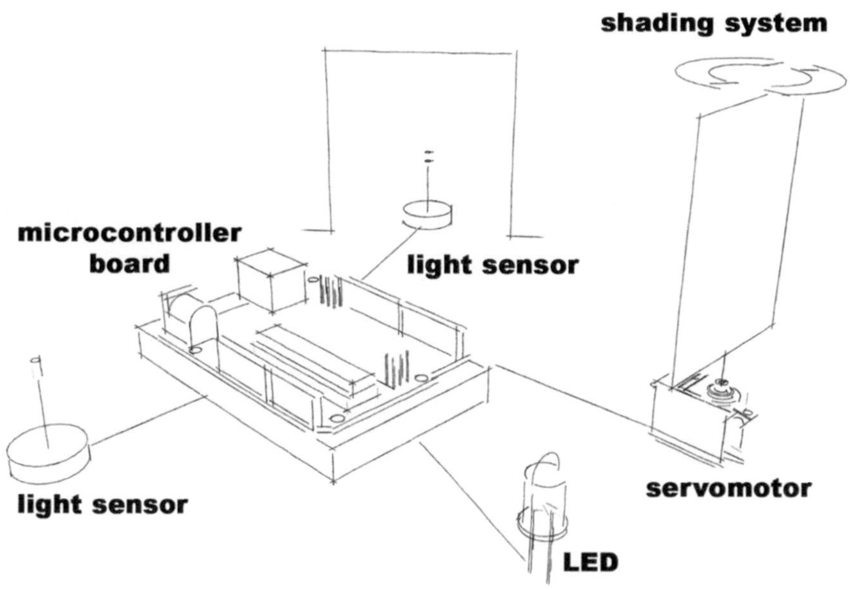

Fig. 2.2 Schematic representation of the chosen hardware components of the prototype

2.2.1 Hardware Components

As shown in Fig. 2.2, the hardware components of the system are:

two light sensors, located inside and outside the demonstrator respectively;
the LED artificial lighting system;
a servomotor connected to the mobile screen;
a microcontroller board;
connections and resistances.

Brightness Sensors
Internal and external brightness is measured by two analogue ambient light sensors distributed by the *DFROBOT*.

> *DFROBOT Analog ambient light sensor*
> Power supply voltage: 3–5 V
> Interface: analogue
> Technology: Photoresistor, returns a voltage value (direct current) based on the amount of light that reaches the instrument. High brightness corresponds to lower values.

The reading and writing list of the value on the serial port previously opened in the program is as follows:

```
void sensor_brightness( )
{
    int light;
    light=analogRead(n°pin);//connects light to the analogue pin of the sensor
    Serial.println(light,DEC);//prints the light value on the serial port
    Delay(1000);
}
```

It is, however, necessary for applications which require the translation of values from millivolt to lux, to provide the appropriate conversion equation.

It is possible to use sensors with higher accuracy, but higher costs, which could be used in specific applications in real cases. For example, the sensor produced by TAOS inc. and distributed by *Adafruit* has both an infrared diode and a full-spectrum diode.

TSL2561 digital brightness sensor
Power supply voltage: 2.7–3.6 V
Operating temperature: −30 to +80 °C
Dynamic operating range: 0.1 to 40,000 lx

The TSL2561 is supplied assembled but it is necessary to solder (e.g. Jepson et al. 2012) a 6-pin *pin header in* order to make the connections easily. The software listed is more articulated than the DFRobot sensor and there are specific libraries developed by Adafruit for its sensors, which allow you to directly obtain the output value in lux.

LED Lighting

The lighting system consists of a 10 mm RGB LED with diffused light produced by CHINA YUN SUN LED TECHNOLOGY CO. The absorbed current intensity is 20 mA, while the average light intensity is 550 mcd for the red channel, 250 mcd for the green channel and 700 mcd for the blue channel. The reference voltage varies depending on the colour, as in the case of the different monochromatic LEDs, and has an average value of 2.0 V for red, 3.2 V for green and 3.1 V for blue. In order to be able to guarantee adequate voltages without exceeding the maximum threshold indicated by the manufacturer, three resistors were used: 330 Ω for red and 180 Ω for green and blue. Table 2.1 shows the resistances for different types of LEDs according to the power supply. Table 2.2 shows the specifications of the RGB LED used.

Table 2.1 Resistors to be assigned to the different types of LEDs based on the LED colour and power supply specifications provided by the pin to which they are connected

Led colour	Forward voltage (V)	Resistor for the power pin	
		20 mA and 5 V (Ω)	10 mA and 5 V (Ω)
infrared	1.5	180	390
Red	2.0	150	330
Emerald (green)	3.3	100	180
Blue	3.3	100	180
White	3.3	100	180
UV	3.3	100	180
Orange	2.0	150	330
Yellow	2.1	150	330
Green Yellow	2.2	150	330

The values are to be intended as indicative and must be verified on the basis of the power supply characteristics requested by the manufacturer (Majocchi 2011)

Table 2.2 Operating voltages typical of the type of RGB LED with diffused light used in the example

LED colour	Forward current (mA)	Forward voltage (minimal) (V)	Forward voltage (typical) (V)	Forward voltage (max) (V)
Red	20	1.8	2.0	2.2
Green	20	3.0	3.2	3.4
Blue	20	2.9	3.1	3.3

The features are derived from the technical specifications provided by the manufacturer http://dlnmh9ip6v2uc.cloudfront.net/datasheets/Components/General/YSL-R1047CR4G3BW-F8.pdf viewed 22 December 2013 (checked 1 April 2019)

Servomotor

The shielding component is driven by a GS-9081 micro-servo manufactured by GOTECK, which belongs to the category of servomotors. These motors are designed for modelling applications and are, therefore, characterised by low voltage power supplies compatible with the Arduino board without resorting to additional boards, such as the Arduino *motor shield* board, which allows an external power supply at higher voltages. The servomotors also have absolute position programming, which allows the achievement of high precision. The selected motor allows a rotation of 180° with a central point at 90°, like most servomotors on the market. The Arduino team has developed a special library to facilitate the management of servo motors at software level, which allows the definition of "servo" variables, the reference pins "servo.attach(pin)", the motor position in degrees "servo.write(degrees)" and in absolute values in microseconds "servo.writeMircoseconds(…)". The library also enables the reading of values in position degrees at software level and not directly from the motor "servo.read( )", to check whether the motor is assigned to a pin "servo.attached( )" and, lastly, to disconnect a servo from a pin "servo.detach( )", mainly used for reasons related to the activation of the PWM from digital pins 9 and 10. Specifically, the characteristics of the servomotor used, derived from the manufacturer's board (see the Goteck site), are:

GS-9081 microservo
Weight: 9 g
Dimensions: 23 × 12.5 × 30.0 mm
Operating speed: 60° in 0.10 s at 6 V—0.12 s at 4.8 V
Maximum load: 1.5 kg-cm at 6 V—1.1 kg-cm at 4.8 V
Voltage: 4.8–6 V

Microcontroller Board

The simulation is carried out using an Arduino *Uno R3* board, based on the ATMEGA328 microcontroller (see also Evans 2011; Kelly and Timmis 2013; Margolis 2012; Timmis 2011; Schmidtg 2011 for early support). This board has 14 digital input/output pins, six of which can be used as PWM output, and six analogue input pins. The operating voltage is 5 V, the recommended external power supply is 7–12 V within 6–20 V limits. More information and the entire hardware scheme of the board can be found at https://store.arduino.cc/arduino-uno-rev3 (December 2018).

The Hardware Diagram Developed

The schematic definition of the hardware components shown in Fig. 2.2 and the specifications of the single hardware components used allowed the development of the final design scheme of the demonstrator. Figures 2.3 and 2.4 illustrate the connections between the components on the basis of a visual representation, which reproduces the hardware layout of the demonstrator (*breadboard* view) and the hardware scheme. Processing is carried out using the free *Fritzing* software versions 0.7.11 and 0.8.5 (current version is 0.9.3b).

The diagram enables the reading of the connections between the different hardware components. The black connections indicate the links with the GND pins (mass of the board—intended as 0 V), while those in medium grey in the figure (generally marked in red) indicate the connections with the 5 V pin used to power the sensors and the servomotor. The red LED is powered by pin 13 and is switched on and off by setting the output values of the terminal as HIGH (5 V) and LOW (0 V). The RGB LED is powered by the outputs of digital pins 9, 10 and 11 according to a PWM signal that varies between 0 and 5 V. Remember that it is necessary to interpose resistances to

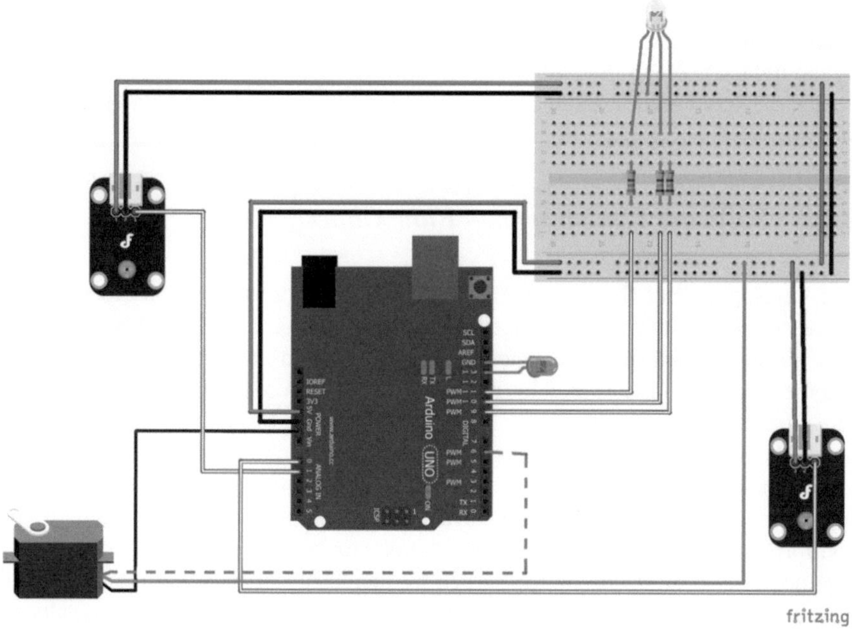

Fig. 2.3 Breadboard view of the hardware system used for the prototype generated using the open source Fritzing

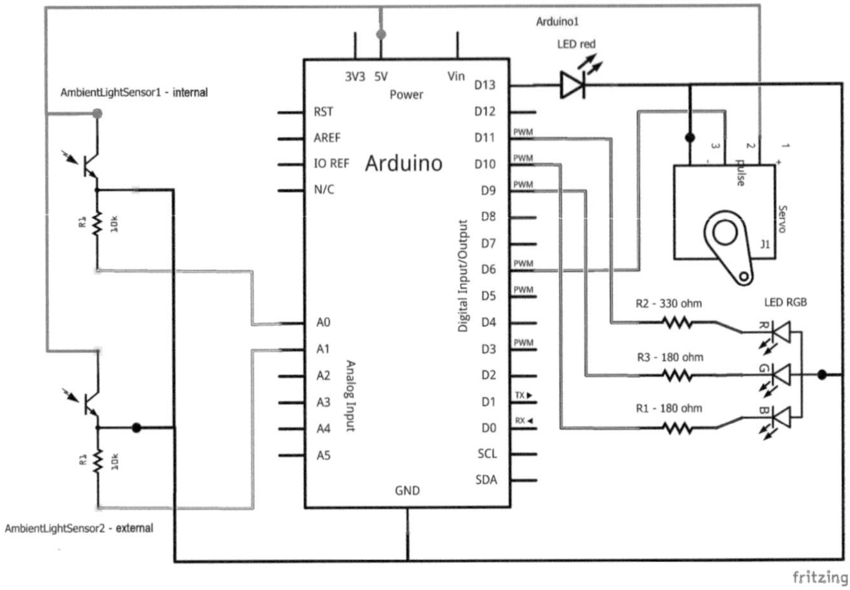

Fig. 2.4 Hardware scheme of the prototype generated by the open source Fritzing

avoid exceeding the maximum voltage of the individual colour channels of the LED. The connections made with the double grey line (analogue pins—on the left of the board) of the diagram in Fig. 2.3. characterise the transmission of the values read by the light sensors to the Arduino board through analogue input pins A0 and A1. Lastly, the servomotor reads the setting position from digital pin 6, via PWM signal.

2.2.2 *Software Components—Flow Chart*

The development of the demonstrator software and hardware is preceded by the processing of the flow chart of operation shown in Fig. 2.5. Value A refers to the light sensor located inside the room, value B refers to the sensor located outside and the curtain and LED variables refer to the operation of the artificial lighting and screen management systems. The logic flow reads the brightness value measured inside the environment and compares it with an acceptability range. If A conforms to the range, the end of the procedure is reached, while if it is inconsistent with the reference, other operations are carried out to obtain the brightness value in the comfort range. The reading of the second probe makes it possible to check whether it is possible to intervene on the screen by increasing or reducing the amount of natural light entering. If the reading of B is not favourable or if the threshold value of the screen has already been reached, the artificial light is switched on or off. There

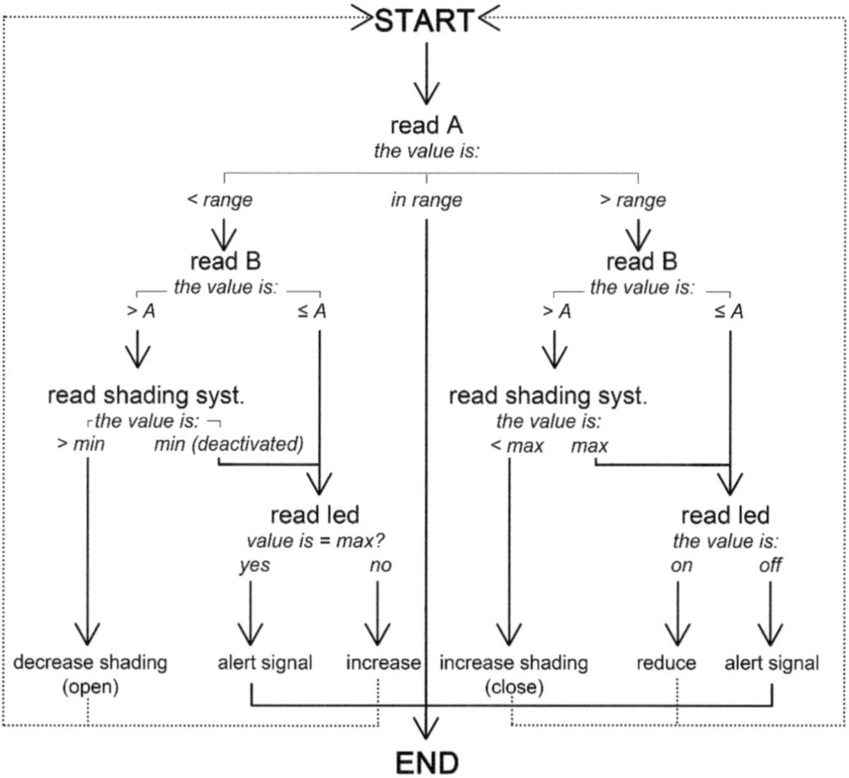

Fig. 2.5 Flowchart of the prototype control system. It was elaborated before the hardware scheme to define the general workflow of the system

are control points that allow you to exit the procedure when the conditions do not guarantee the achievement of the reference range.

The translation from the flow chart to the program list has made some changes necessary in order to:

- organise the logical flow into separate procedures to streamline the main list. The organisation into procedures also allows greater control and management of the logical flow of information;
- relate the operation to the actual reading of the brightness sensors used, which proved to be inversely proportional to the amount of light (deviating from what is defined in the flowchart where the reading is intended as the amount of lux);
- introduce a number of measures to increase the stability of the system. It is important, especially in intermediate situations, to avoid the continuous switching on and off of the LED and repeated movements of the screen;
- allow the LEDs to be operated only when the natural light is not actually sufficient. The decision was therefore made to maximise natural light. Another solution is to modify the demonstrator by introducing, for example, a temperature sensor that

allows you to choose whether to maximise the natural component or minimise solar contributions;
- close the screen when the external brightness value falls below the threshold value, to ensure occupants' privacy during the night.

2.2.3 Software Components—Script

The software list has been developed in the Arduino environment using version 1.0.3 of the open source software for programming Arduino boards based on *Processing* (Reas and Fry 2010), which can be downloaded free of charge from the Arduino.cc website (Banzi et al.)—the current version is 1.8.8 (https://www.arduino.cc/en/Main/Software, viewed December 2018).

The *servo.h* library was used to prepare the software to allow easy management of the servomotor.

The development of the list includes the definition of the main variables, the setup and the main loop.

Three variables for managing the servomotor:
- variable servo type;
- *integer* variable, which defines the angle in degrees;
- additional integer variable used to increase system stability and reduce noise from continuous servo repositioning to the same values;

Nine variables for the management of the RGB LED:

- three integer variables for the management of the three RGB channels of the LED and the assignment of the digital pins;
- integer variable for the value to be assigned to the three channels;
- additional integer variable to increase system stability and reduce flicker;
- two variables for light sensors
- two integer variables to which the two values read by the brightness sensors will be assigned;

Two variables for the comfort range:

- integer variable that defines the central value of the comfort threshold;
- integer variable that defines the width of the comfort zone around the threshold value.

The setup, i.e. the program setting, is useful to manage the opening of the serial port necessary to communicate with any computer connected to the device, such as system operation reports, identification of any anomalies, writing operating values. The same serial port is used for communication with Rhinoceros. The setup procedure includes the definition of the output pins for the management of the servomotor, the output pins for the management of the red control LED and the three pins connected to the RGB LED. Lastly, the comfort range is defined in terms of threshold and amplitude and the variables used are set to zero values to allow the launch of the body of the program. The body of the program is organised through separate procedures, according to an approach that facilitates the understanding of the logical flow, allows subsequent intervention on the list with greater ease and streamlines the main loop. In the Arduino software, procedures can be implemented using secondary loops. Nine secondary procedures have been developed in this case:

1. reduction of brightness;
2. increase in brightness;
3. reduction of the LED;
4. increase in the LED;
5. reduction of the incoming natural light—closing of the screens;
6. increasing the incoming natural light—opening the screens;
7. variable setting;
8. energy saving for the values compliant with the range assigned;
9. verification of operation and serial communications.

The last procedure manages the operation of the red LED for signalling possible anomalies and allows communication with the computer, if it is connected to the Arduino Uno board. This last secondary loop is then modified to communicate with the Rhinoceros software via the *Grasshopper* and *Firefly* plug-ins, using the serial port.

The main loop refers directly to procedures 7 and 9 and, based on the values read by the internal light sensor, then activates a procedure of your choice between 1, 2 or 8. The other procedures are referred to within the secondary loops depending on the logical flows.

The variations induced on the position of the screen and on the brightness of the LED have been set according to predefined steps of 15° for the screen and 64 values for the LED, considering the 256 permissions of the 8 bits of the digital pins. This choice is dictated by the need to reduce fluctuations around the threshold value and increase system stability (Fig. 2.6).

The latest version of the software list developed for the management of the demonstrator is shown at the end of the chapter.

2.3 Connection with a Parametric CAD Software—Real-Virtual Interface

The phase after the development of the demonstrator envisages the connection of the home automation system with the CAD Rhinoceros software, version 5, release SR6 developed by *Robert McNeel & Associates.* The practical demonstration of the feasibility of the connection makes it possible to understand new design and maintenance

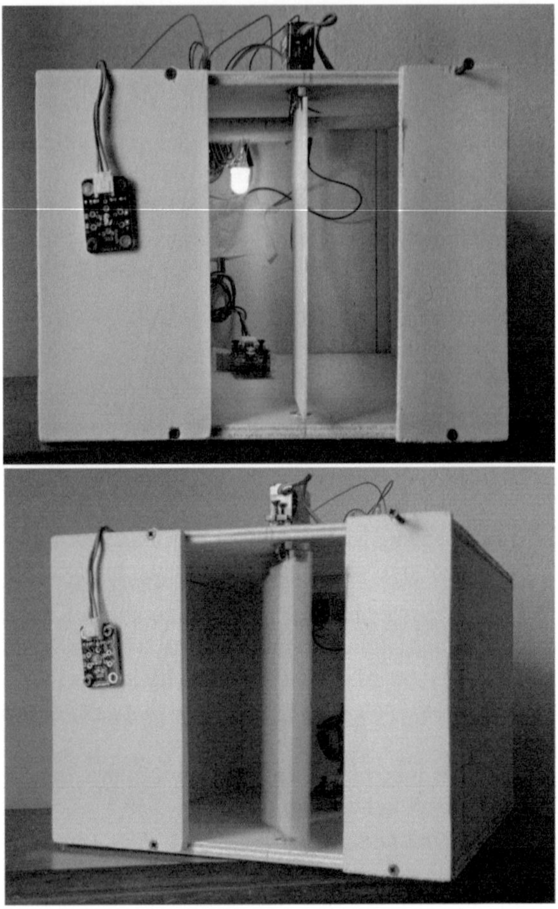

Fig. 2.6 Different configurations of the prototype according to environmental conditions

scenarios for the architecture and, in addition, the smart management of resources and systems.

2.3.1 Connecting Modalities

The organisation of the flow of information of the demonstrator can be represented according to the diagram shown in Fig. 2.7. The operation of the sensor/actuator system is independent of the connection to the computer and can be powered by a USB port (computer, USB power supply for mobile phones, …) or other external source in compliance with the voltage limits of the board. The sensors read the value of a specific variable and communicate it to the board via an electrical input signal. Then, after processing the information, the microcontroller sends a response to the

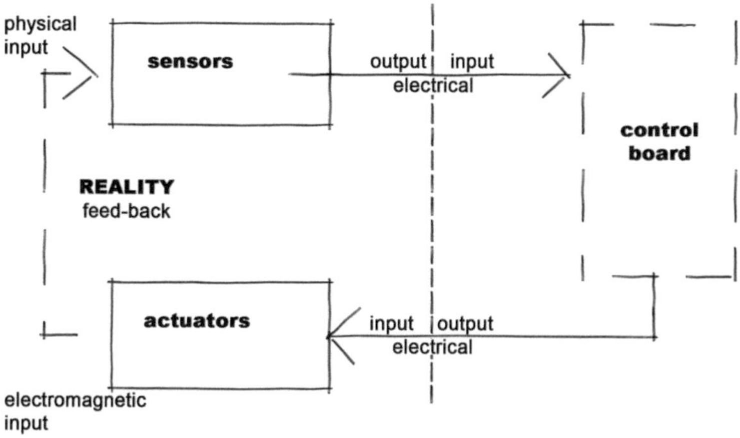

Fig. 2.7 Schematic representation of the workflow scheme of the prototype or of a general sensor/actuator system

actuators in compliance with a feedback cycle. Potentially, additional controllers retroaction may be included to optimize reactions (e.g. Åström and Hägglund 1995).

The connection with CAD Rhinoceros software has been developed by interfacing with the Arduino board, which assumes the role of interface between the real world and the virtual world. Figure 2.8 shows a simplified diagram of the information flow. In the example of the demonstrator, the flow is one-way from the real world to the virtual one but, as reported by numerous examples in literature (the projects reported on the *Firefly Experiments* website, the research of the MIT *Media Lab* and the *Senseable City Lab*, the projects reported in Hwang et al. (2006) and in Imperiale (2001) and the installations of *kinetic architecture*), it is possible to develop reverse and bidirectional flows. Bidirectional flows can be used to manually intervene on a home automation system bypassing the automatisms in order to respond to specific user needs. Reverse flows allow the driving of actuators from the virtual model, by conforming real geometries on the basis of virtual geometries, for example, or by managing numerically controlled machines on the basis of paths drawn in a digital environment. The virtual model can also be optimised in relation to simulations and evaluations carried out through dedicated programs and the results can be translated into the real world using actuators.

The transmission of data from the real world to the virtual world is carried out through platforms for the collection, sorting and translation of information according to a flow outlined in Fig. 2.9. Physical reality is read and translated by sensors and modified by actuators. The Arduino board is a platform capable of receiving raw or processed data from the real world, using it and managing the actuators and transmitting them to other platforms or storage systems in different ways. The data transmitted or stored can be read and translated by software platforms and subplatforms for use in other environments, such as the virtual world of CAD models. Geometric models and parameters can also be transmitted and processed by analysis

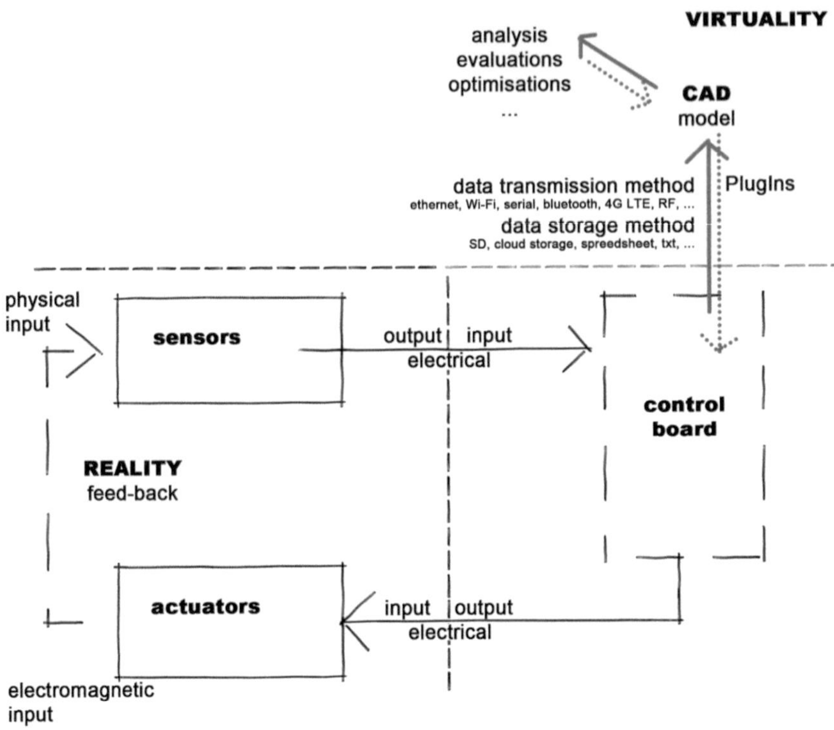

Fig. 2.8 Schematic representation of the workflow between the prototype and the CAD software. Dotted lines represent potential flows not included in the sample demonstration

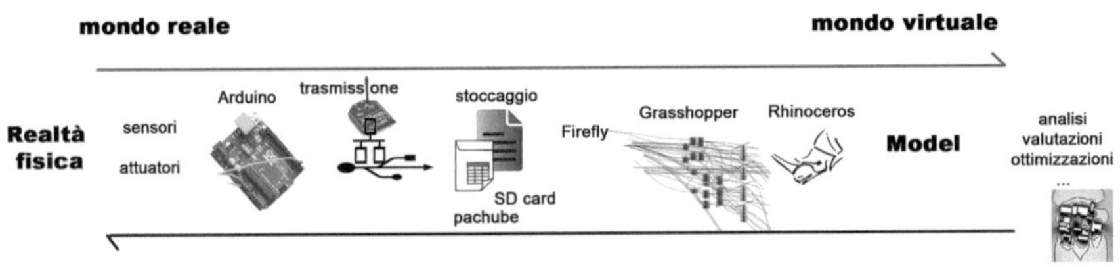

Fig. 2.9 Schematic representation of software connections used for the prototype and potential alternative solutions

and/or optimisation software. This flow can be reversed to allow real-world physical parameters to achieve certain performances in response to specific needs.

2.3.2 Functioning

The connection between the Rhinoceros CAD platform and the Arduino board is programmed via a serial port using the USB connection. The plug-ins required for the connection are Grasshopper® and Firefly.

Grasshopper® is a graphical algorithm editor firmly integrated with the Rhinoceros 3-D modelling tool (see also Payne and Issa 2009). The copyright® of the plug-in is owned by Robert McNeel and Associates and is distributed free of charge. The version used during the demonstration is 0.9.0064.

Firefly, plug-in born in 2010, is a set of tools built for use within Grasshopper®. The development of the tools is due to continuous updating thanks to numerous contributions by experts and researchers on an international scale, coordinated by the two founders of Firefly: Andy Payne of *Harvard GSB* (Boston), and Jason Kelly Johnson of the *California College of the Arts* (San Francisco).

Communication between Rhinoceros and the Arduino board is carried out via a serial port, although it would be possible to use a different IP address or storage and transmission system. Secondary procedure 9, introduced in point 2.2.3, has been modified to transmit the data necessary for the real-time implementation of the two variables connected to the virtual CAD model to the serial port. The related variables are:

- the position of the screen, driven in the real world by the servomotor in compliance with the demonstrator management procedures described above;
- the switching on of the artificial LED lighting, assumed as an ON–OFF value, differing from the actual operation which envisages intermediate steps.

In order to streamline the script in Grasshopper, the decision was made to transmit via serial port only one value, capable of synthesising both the angular position of the screen and the ignition of the LED system. The position of the screen is established by an angle of rotation around an axis and, as such, 360° can be added without changing its positioning at geometric level. On the basis of these premises, the value of the angle transmitted on the serial port has been modified at software level to increase the value transmitted by a rotation angle of 360° if the LED is on. The status of the LEDs can therefore be obtained by means of an integer division with a divider of 360.

The script, shown in Fig. 2.10, is divided into three macro parts:

- the first, located on the left in the figure, includes the procedures to identify the serial ports used by the computer, to define the one used for the connection between the computer and Arduino (the computer assigns a fixed COM for each Arduino board), choose the baud value of operation (speed expressed in [bit/s]), which must coincide with that indicated in the software list, and identify the refresh reading time. In addition, a logical variable has been added to define the switching on and off of the connection;
- the second, central position in the figure, reads the data transmitted on the serial and uses it to rotate the geometry of the screen within the CAD model;

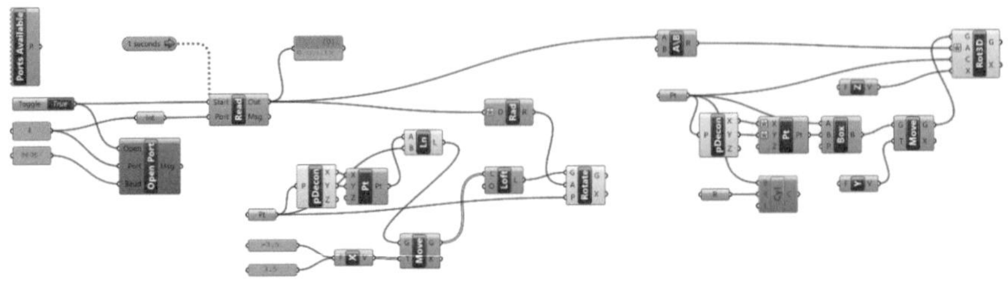

Fig. 2.10 The developed Grasshopper script to connect the Arduino board (see Fig. 2.11) through serial port and the developed CAD model in Rhinoceros®—see Fig. 2.12

– the third, on the right in the figure, reads the data transmitted, interprets it through the whole division, and rotates the indicator in the CAD model, identifying the switching on or off of the LED.

The Grasshopper plug-in allows the representation within Rhinoceros geometries that can be modified in real time. In the case of the demonstrator, the static model shown in Fig. 2.12 has been recreated in Rhinoceros, with the addition of the elements drawn in Grasshopper which are capable of modifying themselves in real time.

Figures 2.13 and 2.14 illustrate the functioning of the real and virtual demonstrators, demonstrating the functioning of the bridge between the real world and the simulated virtual world.

Fig. 2.11 Detail photograph of some physical components of the prototype. It is possible to see the microcontroller board and the servo motor

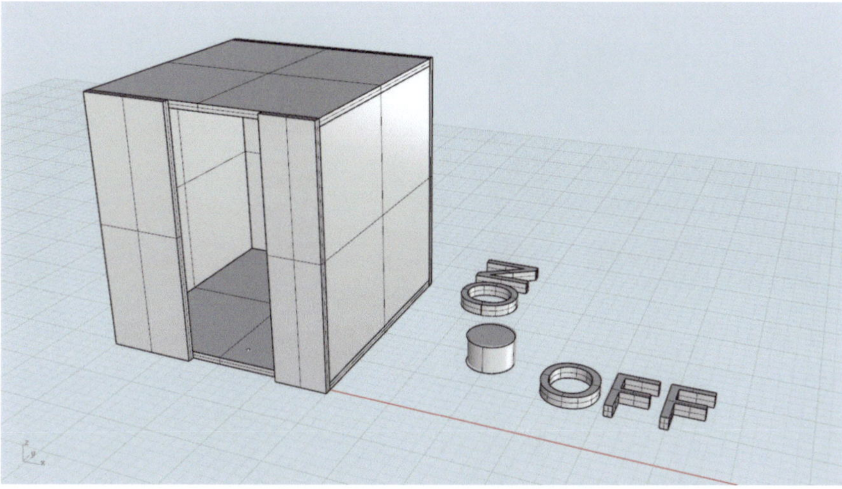

Fig. 2.12 The static part of the CAD model implemented in Rhinoceros® 5 SR6

Fig. 2.13 Sample of the virtual (CAD)-real world connection (night hours—shading activated and led activated)

Fig. 2.14 Sample of the virtual (CAD)—real world connection (day time, shading open and led deactivated since natural light is sufficient to reach the fixed internal threshold)

2.4 Discussion

The simulation proposed has shown how it is possible to implement customised sensor/actuator solutions using open source software and hardware. The integration with a CAD platform broadens the scope of application and the possible effects of these technologies by foreshadowing development scenarios in different fields of knowledge. As reported by numerous literature studies (such as Maldonado 2007), advanced technologies and the current implications of ICT are changing the relationship between reality and representations of what is real. At the same time, contemporary techniques of geometric elaboration and tools for the optimisation of design processes, analysis and production of complex shapes and performance evaluation are interfacing with the emerging parametric, algorithmic and generative shape software in a new virtual dimensionality with enormous potential (Woodbury 2010; Ceccato et al. 2010; Terzidis 2006). The way we think about technology is changing radically and the links between the real world and the virtual world, also understood as a geometric model, can only accelerate and increase this change (Braham and Hale 2007).

At least four different levels of interaction can be identified, characterised by specific platforms, management locations, production, storage, sorting, use and data optimisation:

- level 1: interaction between physical reality and stand-alone sensor/actuator system. This point includes numerous applications of home automation, robotics and system control;
- level 2: interaction between physical reality, actuator sensor system and technologies for transmission, storage and visualisation of data and information sent and received. Data storage and visualisation can be used to conduct scientific monitoring, reading and transmission of data and information remotely, from the local network or from the Internet for example, from mobile phones or other devices, populating tables and diagrams;
- level 3: interaction between physical reality and virtual models allowing the visualisation of geometries and/or governance of an actuator system. This point includes maps of variables and data of a geo-referenced and/or geometric nature, opening up to numerous applications. For example, it is possible to report the occurrence of anomalies and failures in real time and locate them in a virtual model of the geometries, helping to manage emergencies and maintenance activities. This also includes many applications of kinetic or adaptive architecture;
- level 4: use of virtual models to evaluate, optimise and consequently modify the performance of a real system through a system of actuators and cycles of real-virtual feedback and simulations. This point is based on the optimisation of performance in response to specific needs, allowing possible self-learning actions based on the progressive increase of feedback data available. It would be possible to envisage a management of the operation of systems and buildings optimised according to boundary conditions and specific needs, inducing defined feedback and/or choices through repeated simulations and progressive evaluations, using genetic optimisation algorithms including *Galapagos*.

Models and data, simulated, stored, transmitted and processed, can interface directly with reality, overlapping it as additional layers, viewable through mobile phones, palmtops, screens and glasses. The theme of augmented reality, although not the central topic of the chapter, is interrelated with all the levels briefly described, assuming considerable importance especially from the second point.

The connection between monitoring, storage and data analysis installations and systems also opens up numerous lines of study and design implementations, both during operation and as regards building maintenance, in order to develop protocols, plan interventions for energy certification in operational rating (e.g. Healy Beng 2011) and reduce actual consumption.

2.5 The Arduino Script

```
/*
Internal lighting management program using RGB LED and shading system control devices.
Elaborated by Giacomo Chiesa.
Version 0.3 – 2 December 2013
*/

// libraries
#include <Servo.h>

//variables
Servo shading;    //shading is defined as a servo motor variable
int shading_degree; //shading_degree is the value assigned to the servo motor
int nonoise;   //variable included in order to avoid reset rumours on the same value

int ledR=9;    //setting LED pins
int ledG=10;
int ledB=11;
int led; //led is the value assigned to LED pins
int novar; //variable defined to avoid LED fluctuations when the same value is set

int ReadA; //internal luminosity reader – it will be assigned to pin A0
int ReadB; //external luminosity reader – it will be assigned to pin A1

int setcomfort; //comfort set point – is the central value in the threshold definition
int rangecomfort; //amplitude of the comfort threshold – inside this domain actuator inputs will not change – stability
interval

//setup
void setup()
{
Serial.begin(9600); //the serial port is set to 9600 baud – this is used for Rhino connection only
shading.attach(6); //digital pin 6 is set as input/output for the servo motor
pinMode (ledR, OUTPUT);//LED pins are defined as outputs
pinMode (ledG, OUTPUT);
pinMode (ledB, OUTPUT);
pinMode (13, OUTPUT);

setcomfort=(700); //comfort threshold definition
rangecomfort=(50);//the amplitude of the comfort threshold is defined
led=0; //LED operational variable (on/off and dimmering) – at starting point it is set to 0
shading_degree=0; //Starting servo position is set to shading condition closed
shading.write(shading_degree); //define the starting shading conditions (servo position 0)
digitalWrite (ledR, led); //define starting LED conditions to all pin
digitalWrite (ledG, led);
digitalWrite (ledB, led);
nonoise=0;
novar=0;
}

//loop main program
void loop()
{

  Pr_intro(); //procedure intro()
  Pr_set_variables(); //procedure for variable setting
  ReadA = analogRead(0);
```

```
 if (ReadA < setcomfort-rangecomfort)
{
  //Serial.println("reduce luminosity");
  Pr_reduce(); //procedure to reduce internal light
  }
else
{
  if (ReadA > setcomfort+rangecomfort)
  {
    // Serial.println("increase luminosity");
    Pr_increase(); //procedure to increase internal light
    }
   else
   {
     Pr_inrange();  //procedure when light is in range
   }
  }
 delay(1000);
}

//PROCEDURE reduce internal light
void Pr_reduce()
{
 ReadB=analogRead(1);
 if (ReadB < ReadA)
 {
  if (shading_degree <=0)
  {
    shading_degree = 0;
    led = 0;
   }
   else
   {
    led = 0;
    shading_degree=(shading_degree - 15);
    if (shading_degree <=0)
    {
      shading_degree = 0;
    }
   }
  }
  else
  {
   if (ReadB > setcomfort+2*rangecomfort)
   {
     shading_degree = 0;
   }
   else
   {
     shading_degree = 90;
   }
   Pr_ledreduction();   //Procedure to reduce LED intensity
   }
}
```

```
//PROCEDURE increase internal light
void Pr_increase()
{
 ReadB=analogRead(1);
 if (ReadB < ReadA)
 {
  if (shading_degree >=90)
  {
   shading_degree = 90;
   Pr_ledincrease(); //Procedure to increase LED intensity
  }
  else
  {
   led = 0;
   Pr_shadereduction();   //Procedure to reduce the shading effect (shading opening)
  }
 }
 else
 {
  if (ReadB > setcomfort+2*rangecomfort)
  {
   shading_degree = 0;
  }
  else
  {
   shading_degree = 90;
  }
  Pr_ledincrease();
  }
}


//PROCEDURE Pr_ledreduction - reduce LED intensity
void Pr_ledreduction()
{
 //Serial.println("led intensity reduction");
 int i=0;
 if (led == 0)
 {
  for (i =0; i<3; i++);
  digitalWrite(13, HIGH);
  delay (2000);
  digitalWrite(13, LOW);
  delay(2000);
 }
 else
 {
  led = (led - 64);
  if (led <= 0)
  {
   led = 0;
  }
 }
}
```

```
//PROCEDURE Pr_ledincrease — increase LED intensity
void Pr_ledincrease()
{
 // Serial.println("increasing LED intensity");
 int i=0;
 if (led == 255)
 {
     for (i =0; i<2; i++);
     digitalWrite(13, HIGH);
     delay (500);
     digitalWrite(13, LOW);
     delay(50);
     digitalWrite(13, HIGH);
     delay (500);
     digitalWrite(13, LOW);
     delay(50);
  }
  else
  {
   led = (led + 64);
   if (led >= 255)
   {
    led = 255;
   }
  }
}

//PROCEDURE Pr_shadeincrease — reduce the entering light (shading activation)
void Pr_shadeincrease()
{
 //Serial.println("activating shading");
 if (shading_degree <= 0)
 {
  shading_degree = 0;
 }
 else
 {
  shading_degree = (shading_degree-15);
  if (shading_degree < 0)
  {
   shading_degree = 0;
  }
 }
}

//PROCEDURE Pr_shadereduction() — increasing the entering light (shading reduction)
void Pr_shadereduction()
{
 // Serial.println("decreasing shading");
 if (shading_degree >= 90)
 {
  shading_degree = 90;
 }
 else
 {
  shading_degree = (shading_degree+15);
```

```
  }
}

//PROCEDURE set_variables variabili
void Pr_set_variables()
{
 if (led == novar) {
 }
 else {
    digitalWrite (ledR, led); //define LED values
    digitalWrite (ledG, led);
    digitalWrite (ledB, led);
    }
 novar = led;
 if (shading_degree == nonoise) {
 }
 else {
      shading.write(shading_degree);
      }
 nonoise = shading_degree;
 delay(500);
}

//PROCEDURE inrange to turn off LED if the outside luminosity increases
void Pr_inrange()
{
 if (led > 0)
 {
  if (ReadB < ReadA)
  {
   led = 0;
  // Serial.println("turn off LED");
  }
  else
  {
  // Serial.println("ok");
  }
 }
 else
 {
 // Serial.println("ok");
 }
}

//PROCEDURE intro to test serial connection state
void Pr_intro()
{
  ReadA = analogRead(0);
  ReadB = analogRead(1);
 // Serial.print(ReadA);
// Serial.print(", ");
 // Serial.print(ReadB);
// Serial.print(", ");
 //Serial.print(led);
// Serial.print(", ");
```

```
if (shading_degree < 0) { shading_degree=0; }
 if (led > 0) {

  Serial.println(shading_degree+360);
 }
 else {
   Serial.println(shading_degree);

 }


 digitalWrite(13, HIGH);
 delay (500);
 digitalWrite(13, LOW);
}
```

References

Åström KJ, Hägglund T (1995) PID controllers: theory, design, and tuning, 2nd edn. Instrument Society of America, Research Triangle Park, North Carolina. http://aiecp.files.wordpress.com/2012/07/1-0-1-k-j-astrom-pid-controllers-theory-design-and-tuning-2ed.pdf, last view Apr 2019

Braham WW, Hale JA (eds) (2007) Rethinking technology. A reader in architectural theory. Routledge, London and New York

Ceccato C et al (eds) (2010) Advances in architectural geometry 2010. Springer, Wien

Evans B (2011) Beginning Arduino programming. Apress, Springer Science distribution, New York

Healy Beng D (2011) Asset ratings and operational ratings—the relationship between different energy certificate types for UK buildings. Ph.D. thesis, University of Cambridge, Cambridge

Hwang I et al (eds) (2006) Verb nature, architecture boogazine. Actar, Barcelona

Imperiale A (2001) Nuove bidimensionalità. Tensioni superficiali nell'architettura digitale. Testo and Immagine, Roma

Jepson B et al (2012) Learn to solder. O'Reilly, Sebastopol

Kelly JF, Timmis H (2013) Arduino adventures. Escape from Gemini Station. Apress, Springer distribution, New York

Majocchi S (2011) Primo passi con Arduino. Elettronica In 162, supplement

Maldonado T (2007) Reale e virtuale, 2nd edn. Feltrinelli, Milano

Margolis M (2012) Arduino cookbook, 2nd edn. O'Reilly, Sebastopol

Payne A, Issa R (2009) Grasshopper primer, 2nd edn—for version 0.6.0007. Last view Apr 2019. http://www.liftarchitects.com/blog/2009/3/25/grasshopper-primer-english-edition

Reas C, Fry B (2010) Getting started with processing. O'Reilly, Sebastopol

Schmidt M (2011) Arduino. A quick-start guide. The Pragmatic Bookshelf, Raleigh and Dallas

Terzidis K (2006) Algorithmic architecture. Elsevier, Architectural Press, Oxford

Timmis H (2011) Practical Arduino engineering. Apress, Springer distribution, New York

Woodbury R (2010) Elements of parametric design. Routledge, London and New York

http://arduino.cc/, last view Apr 2019

http://dlnmh9ip6v2uc.cloudfront.net/datasheets/Components/General/YSL-R1047CR4G3BW-F8.pdf, last view Apr 2019

http://fireflyexperiments.com/, last view Dec 2013

http://www.fritzing.org, last view Apr 2019

https://www.media.mit.edu/, last view Apr 2019

http://senseable.mit.edu/, last view Apr 2019

http://goteckrc.com/, last view Dec 2013

Chapter 3
Scripting and Parametric CAD Modelling for Performance-Driven Design

3.1 Introduction

> Creating architecture means trying to enucleate one of the probable tangible structural solutions relating to the various architectural systems conceivable for man and his associative life, which the architect is called upon to study from time to time, in a meaningful and relational form (rational, as opposed to the natural alternative of balancing systems). It is a kind of research that is organised and implemented within the context of other systems, including operating systems, as a subsystem or as a partial system of the architectural system to which it refers.
>
> Ciribini (1968, p. 18)

Parametric and algorithmic scripts and mass-customization solutions are perfect instruments for applying and innovating the requirement-driven approach which was developed in the 60s and 70s (e.g. Ciribini 1970). This technological method, which is antithetical to the typological approach, allows for performance evaluation and integration during the different stages of the design process (es. Grosso 2005; Chiesa and Grosso 2019b). At the same time, it can guarantee a good flexibility and a high quality of design. It is directly interrelated with the systems approach, which envisages the building process as a structured and interrelated set of problems and systems—e.g. sociological, technological, structural, economic, constructive, functional, environmental, energetic, aesthetic (Bocco and Cavaglià 2008), and permits a beneficial interrelation between early-design (metadesign, building programming, etc.), design development, building processes and the building industry to fit users' and stakeholders' requirements. The support of digital technologies in this design approach may relate to a large spectrum of aspects including the analysis and the optimisation of processes and/or spatial organisations, the managing/control of flux matrices (e.g. resources, energies, people, stocks, …), computational morphogenesis, etc.

This chapter was co-authored by Enrico Casetta.

G. Chiesa, *Technological Paradigms and Digital Eras*,
PoliTO Springer Series, https://doi.org/10.1007/978-3-030-26199-3_3

In a temporally-driven scheme, Fig. 3.1 illustrates the traditional design-building process interrelated with new instruments and solutions. Early-design steps could be directly interrelated with generative/procedural design, allowing feed-back. The same procedural algorithmic approach can also be used to optimise the building operation. In this sense, the possibility to access detailed three-dimensional design representations, including information related to materials, energy variables, etc. is essential, as is suggested by the dissemination of BIM-based projects. The design development step is divided into preliminary design, definitive design and constructive/executive design, which are the standardized phases of an architectural project in Italy. Each of these phases could be implemented with new parametric software using an integrated approach that interrelates shape formation and evaluation in an optimizing procedure. New digital technologies could be used in every phase of building from design to disassembling—see also Chaps. 1 and 5.

This chapter focuses on the connection between algorithmic design approaches and the need-performance driven design approach—derived directly from the Italian "metodo esigenziale-prestazionale", not strongly detached in origin from the performance concept elaborated at Ulm (see also the works of Maldonado), a school where Ciribini, one of the fathers of the Italian approach, taught, participating in the elaboration of the concept in the late '50s (Matteoli 2013). The chapter is divided into two paragraphs: the first (Sect. 3.2) focuses on the performance-driven design

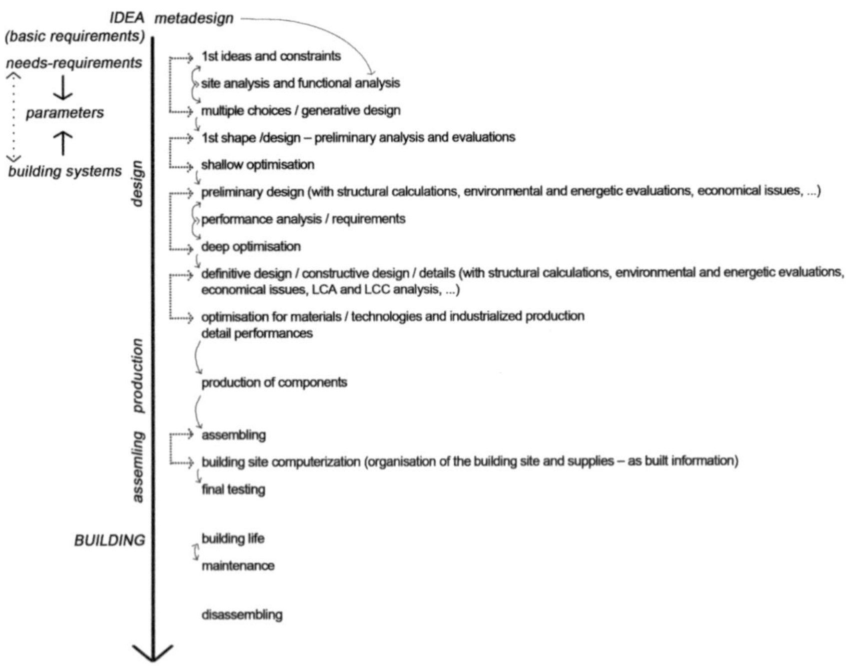

Fig. 3.1 Interconnection between innovative algorithmic technologies/instruments and the architectural process (from design to building demolition)

approach and includes a discussion on how this method, specific to the Architectural Technology field, is based on a perspective that can be directly related to parametric/algorithmic scripting design; the second (Sect. 3.3) tries to detail the potential of this approach by defining two simple applications chosen to be directly related to potential topics in architectural practice but conceived in a way that can help readers understand the potential of this approach rather than giving them a mere list of finished solutions.

3.1.1 Objectives

The main aim of the chapter is to briefly describe how parametric design—especially that of a generative-performative nature—can be directly related to the need-performance-driven approach (see also Sect. 1.1). A design method conceived in the '60s, based on a prodromal vision of what can be achieved with today's instruments. Nevertheless, the chapter is merely an introduction to this complex debate

Chapter 3 also aims (specific objectives) to briefly introduce the need-performance driven approach—based on the Italian Architectural Technology field definition—relating it to the concept of the explicit (vs. implicit) design approach, and to help readers become acquainted with parametric processes—see the specific sample cases presented here. Also in this case, as in Chaps. 2 and 4, the examples are defined to be simple and replicable, without losing their direct link to the need to solve real problems.

3.2 Parametric-Driven Environmental Design

In order to analyse the relationships between need-performance driven design and the possibilities induced by digital media, which are reformulating "the fundamental concepts of the design theory" (Oxman 2006b, 225)—see Chap. 1—it is necessary to remember that the term design is understood here in its broadest sense "to indicate both the activity of analysing the resources available to meet certain needs, and the process of decisions made to achieve a specific goal" (Tosoni 2008, 12). In this sense, as Ciribini reminds us, architecture, from the Greek tektàino, invention and operation at the same time, is configured as a discipline "capable of judging all those works built by the individual arts" (Vitruvius Pollio, 15 BC 2002), or "discipline with tasks and possibilities that transcend the current meaning of science and art" (Ciribini 1968), where "everything serves to fertilise architecture" (Piano 2002). The same Kantian vision links us to the concept of "art of the system", in which "multiple knowledge is gathered under an idea" (Kant 1787). Moreover, acting in its constructive meaning is always related to an environment, stressing how spatial and temporal relationships make the concept of design part of a wider meaning that can be represented by the term design. Where environment is to be understood as an interweaving of systems, in which the act of design is characterised by continuity as it is characterised by "uninterrupted dependence between its nominal distinctions" and "perennial development"—unitary, but under continuous re-examination, analysis and correction, through to rejection in some cases (Ciribini 1968).

The multiple aspects of Kantian-influenced knowledge, must, therefore, find the rational opportunity to address "tangled problems, with imprecise contours" (Tosoni 2008) in the individual areas of knowledge involved in design, allowing the birth of a "science of design", understood as the science of the artificial (Simon 1981), "aimed at decision, organisation, planning and control" specifying that "in the design field, knowing is always knowing how to do something, knowing is knowing ways of operating (Tosoni 2008, 17).

This concept is particularly important in the age of IT, where the capabilities of the individual designer can be overcome by the use of integrated systems aimed at managing shared designs generated by the interaction of people with different skills, each of which, in order to be really involved, must have a strong and structured knowledge without which the sharing and quality of the design might not be achieved (Carrara et al. 2014). The unconscious and non-rational use of the tool risks generating repetitive, delocalised designs (and consequently constructions), lacking sufficient technical-cultural elaboration. This lack of technical understanding deprives the figure of the designer of meaning, relegating it to the same level as the banal draughtsman, involved in the world of mass production, delegated the performance of actions such as "copy, paste, mirror" (see also Sennet 2008) and incapable of real design actions. This process risks reducing "residences to mere consumer products […], a monothematic script capable of housing clearly defined lifestyles and working methods, short-term industrial processes and a stream of commercial businesses that are constantly replacing one another" (Droege 2006). In this sense, we must also understand the reference to Celento in Chap. 1, concerning the need to encourage the training of the professional architect, capable of managing digital innovation, of triggering the necessary adaptation of this sphere, overcoming the current problems summed up in the paradigms "forcing a square peg into a round hole"—tool-traditional process deviation—and "horseless carriage"—lack of vision and operating methods—(Kalay 2006). In this background, there is an evident need for method, a changing stimulus and not a static structure, which "acts as a guiding tool in the continuous verification of the models of the components and of the whole" in the act of design for the identification of *coherences* and *homogeneity*, logical on the one hand—structuring and verification, universally communicable in that they can be explained—and intuitive on the other—formal, implicit, infra-subjective (Ciribini 1968). In the act of digital design, especially when it becomes parametric/procedural, there is an evident need to explain methodologies and models, because the implicit action of the traditional design vision, inherent in the action of the designer himself, but not conceived to be communicated or translated into a structural process or even into a logical-linguistic system, other than in the final outcome, is not adaptable to the potential of the medium. The difficulty in moving from an implicit to an explicit method, including the possibility of activating feedback cycles based on the clear definition of objectives and requirements to which the project must respond performatively, is one of the causes of that sort of obsolescence of the figure of the designer identified by Celento (2007). So if the aim of the practice of design is always a vision, a solution to a given problem—which includes methods, organisations, processes, knowledge and cultures (Kalay 2006)—this vision can derive from an explicit or implicit process (Chiesa 2016). However, in the latter case, the designer knows exactly how to resolve it, but doesn't have such a clear idea of the system of relationships, the model and the structure that generates it. In the digital sphere, this lack of explanation leads to a lack of communication, without which the operator and the tool risk being unable to interact correctly. The concept of structure, system of relations and model are articulated in the double concept of Vitruvius's *firmitas*, and of dynamism, of entelechy, i.e.: of orientation/development with respect to an internal final cause, in Aristotle's vision, or as a "centre of energy" characterised by an autonomous development towards an end in an environment in which a unitary complex can be identified, in the Leibnizian vision. In this sense, the theory of the guide, pursued by the designer, is linked to the possibility of making the action—which, as in entelechy, tends to run in one direction—connecting the system to the design in a language changed by cybernetics, effective (Ciribini 1968). The need to possess the necessary knowledge to manage explicitly performing and not only morphological-prescriptive processes, in order to avoid their distorted or reduced application, is also mentioned in Carrara et al. (2014). Regarding the matter of the terms "implicit" and "explicit", see also (Ferraris 2003; Conole and Wills 2013; Woodbury 2010; Reigeluth and Carr-Chellman 2009; Rothenberg 1989).

In this perspective, the esigenziale-prestazionale (need-performance-driven) approach can provide a methodological framework on which to focus the process of re-design and re-engineering of the design practice allowed by what has been called the fifth long wave of technological development (Atkinson 2004; Freeman and Perez 1988; Mansfield 1983), i.e. the new ICT technologies and the consequent algorithmic-procedural effects on design. This approach is also linked to the sixth prodromal wave of change described as "the post-information technological revolution" by Šmihula (2009), based on sustainability, systemic design—and here the link becomes evident—biomimicry, green chemistry, industrial ecology, green nanotechnology, etc.

Section 3.2.1 focuses on the performance-driven approach trying to briefly outline some of its peculiarities related to the digital world, highlighted from the very first theoretical studies, such as the previously mentioned relationship with cybernetics and the direct link to the design approach *by programs*.

3.2.1 Performance-Driven Approach

This short introduction to the performance-driven approach is closely linked to the subject of technology, from the Greek τέχνη—techne "art, skill, cunning of hand"—and λογία—logia—which is "the science of means useful to produce what is needed by a society" (Abbagnano 1961), or also on the one hand "the practical use of science to perform a useful function, [and on the other hand] engineering sciences and their application in industry" (Davies and Jokiniemi 2008). The subject of the need-performance-driven approach is linked to the design action that tends towards the creation, both for the individual and collective society, of a suitable environment in which to live (Torricelli et al 2001). Within the architectural sphere, the technological concept is expressed by the Architectural Technology field, which can also be defined as "the ability to analyse, synthetize, and evaluate building design factors in order to produce efficient and effective technical design solutions that satisfy performance, production and procurement criteria" (School of the Built and Natural Environment, Newcastle City Campus). In this sense, technology, in architecture, is not only a set of tools, but also methods—there is not only one way to make things. Furthermore, as mentioned before, architectural art is a systemic art, comprising a unitary process of ideation, operation and fruition "which is expressed in an environment and on an environment", as mentioned fifty years ago by the technologists of architecture (Ciribini 1968). Moreover, this process is articulated in a vision in which the world can be considered as made up of patterns.

Using the same source (pp. 28–29), it is therefore possible to define—by linking architecture and systemic science—the main terms that contribute to the architectural process.

> *Action* - understood as design action, "can be defined as a process of information-decision" (Bosia 2013). It incorporates the idea of guidance (direction/command) and in this sense owes part of its definition to cybernetic theory ("the art of the pilot or helmsman") - "the command (…) is nothing more than sending messages that modify the receiver's behaviour" (Wiener 1950). This definition should not be confused with an outdated vision of progress, as this concept and its potential are now being constantly enriched thanks to the impact of ICT on the processes and activities introduced in Chap. 1;
>
> *Domain* - the set of circumstances associated with an action;
>
> *Environment* - what is not part of the system specific to the action;
>
> *Around or context* - region of the environment from which (input) and on which (output) a system can undergo or produce actions - also definable as "informer entrance with respect to the system" (Ciribini 1968);
>
> *Function* - the relational framework that the system establishes between input and output, linked to the concept of effectiveness.

In this concept, the structure—which is connected with the concept of structural method here—is the way in which elements and attributes interact in a system united in form. Where form—part of the world over which we have the possibility of control—represents a solution to the problem defined by the context which is, in turn, the part of the world that establishes the need-based framework of form (Alexander 1964, 1965; Ciribini 1968). In this sense, *form* is linked to the concept of *model*, *pattern*, in a multi-language vision of the design process. This process, aimed at finding a solution, does not imply a linear action, but is based on feedback, circuits or slots, according to a performance-driven approach.

The performance-driven approach (metodo esigenziale prestazionale) to building design is based on the idea that a specific behaviour (performance) is required of a given object (building), without focusing on the technology that will be used to perform it. It is, in this sense, opposed to the object-driven approach (traditional vision) in which a given object (building) is described according to the specific technological characteristics that will be chosen for its construction (including materials, shapes, etc.). For this reason, a chair is an element capable of allowing the user to sit, including all related aspects to make this activity possible, including safety and comfort. Quality is, in fact, given by defining WHAT is expected from the building and not by HOW the building is created. The more the building performs in compliance with expectations, the higher its quality. The performance-driven method is, consequently, focused on users. It deals with the definition of *activities* and correlated *needs*. Furthermore, needs are translated into correlated *requirements* that,—see Fig. 3.2—concern both the environmental system and the technological system. Requirements must be checked, through feedback, analysing the *performances* of the in-progress object definition. This process is, therefore, based on a *user—Activities* → *Needs* → *Requirements* ← *Performances* flow, in which feedback activates the continuous process related to the requirement-performance check. This approach has also been defined in several standards.

According to UNI 10838:1999—which replaces UNI 7867-1:1978, UNI 7867-2:1978, UNI 7867-3:1978, UNI 7867-4:1979—it is possible to briefly define these concepts:

> *Activities* of users are acts or actions, by the end users of a building, which require space (e.g. sleeping, relaxing, being ill, eating, reading, etc.);
>
> *Needs* are the expression of requests needed by a user to perform an activity or a technical function;
>
> *Requirements* are technical transpositions of needs—see in particular UNI 8290-2:1983;
>
> *Performances* are technical responses to requirements.

where the term *technical function* refers to any characteristics allowing the achievement of a performance.

Furthermore, needs can be organised according to specific classes in order to structure/help their identification. According to UNI 8289:1981, seven classes of needs can be identified: Safety of users, Wellness/Comfort, Usability, Appearance, Management, Ability to be integrable, Environmental protection. This said, different classifications were defined in further studies based on the environmental and

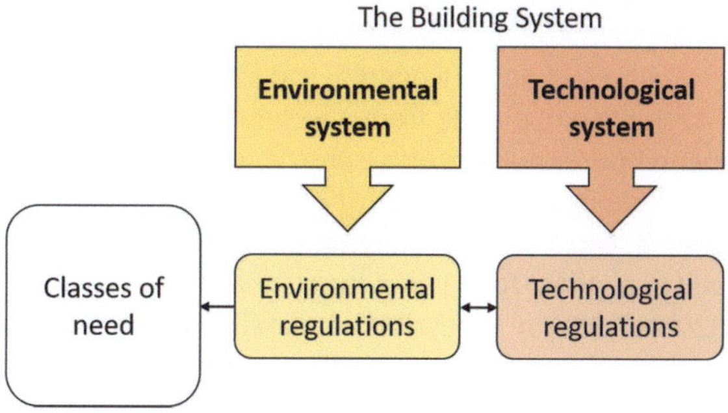

Fig. 3.2 The building system according to the performance-driven approach

sustainable approach to design. UNI 11277:2008 (now withdrawn), focused on eco-compatibility and the consideration of needs-requirements using LC phases, identified three classes of needs: Environmental safeguard, Rational use of resources (a new class), and Comfort/Hygiene/Health.

Concerning requirements, UNI 8290-2:1983 introduces a tentative list of requirements. This standard identifies 63 requirements, including: 1. Reliability, 13. Control of the solar factor, 21. Flowrate control, 39. Thermal insulation, 48. Fire resistance, 63. Ventilation. According to UNI 10838:1999 it is also possible to classify them into seven main aspects: functional-spatial requirements, environmental requirements, technological requirements, operative requirements, durability requirements, and maintenance requirements.

A building expresses its functions via its components, whether it is considered as a combination of technical elements or of material elements (Torricelli et al 2001). Where every component (technical elements)—that is part of the building system and of other sub-systems—helps with several functions. Building functions are interrelated and interdependent, so technological choices acting on a single function may influence other functions (Allen 2005; Reid 1988). For example, acting on airflow control—e.g. for a controlled natural ventilation system or referring to air infiltration—may influence the function of preventing the entry of water, fire control, energy/thermal control, temperature control, humidity control, the supply of fresh air, noise control, the control of comfort in general, maintenance, etc. Nevertheless, in order to study buildings, they have to be organised (divided), highlighting their essential functions and components without losing the systemic vision behind them. Buildings are sets of spatial and technical elements characterised by their functions and relationships. According to the systemic approach, it is possible to identify two interconnected systems: the environmental system and the technological system.

The checking of requirements and performances is generally based on specific indicators, where the indicator is a quantitative or qualitative parameter used to measure the level of performances of a spatial element or technical element, or even of the technical building system. Furthermore, as mentioned earlier, needs and requirements refer to both the *environmental system* and the *technological system*. The first concerns *environmental units*, homogeneous groups of activities which are temporally and spatially compatible. Environmental units are made up of *environmental elements*, which are also defined as *spatial elements*, being related to spatial definition (e.g. living rooms, bedrooms, etc.), that can be defined as part of a space where related activities can be performed. The second—technological—system deals with *technological units* (technological sub-systems), homogeneous groups of functions which are compatible from a technological point of view and are used for environmental performances (e.g. substructures, vertical closure, etc.). Technological units are made up of *technological elements*, components of a specific technological subsystem, capable of partially or totally fulfilling a function related to a technological unit (e.g. direct foundation structures, exterior vertical walls, etc.). According to UNI 8290-1:1981, it is possible to classify technological units in at least eight classes: load bearing structures, envelope, internal partitions, external partitions, building services, security system, internal elements, external elements.

This method can be perfectly adapted to the environmental design approach, especially considering early-design phases—see also descriptions and methods reported in Refs. (Chiesa and Grosso 2017a; Grosso 2005). Furthermore, this design method is perfectly adaptable to the concept of parametrisation and algorithmic approach to building design, being based on feedback cycles.

This concept has been highlighted since its historical definition, being directly inter-connectable also with the concept of industrialisation. This approach can be devoted to "instruments and procedures to define building systems" and this process can be conceived using a program approach (Cavaglià et al. 1975). The identification of activities, needs, related spatial and technological requirements is also directly connected with the further need to define matrixes of flows, that include movements of people, goods, materials etc. All these aspects, and particularly the optimisation of requirements-performances and the analysis/organisation/optimisation of flows, can be at least partially solved using algorithmic design. These aspects cannot, however, be considered as a fixed homologated process, as users and their needs are not univocal, but are based on a very large range of behaviours, which further change during time, including variations due to type of activity, age, cultural background, etc. All these factors have to be assumed in the definition of a dedicated series of models assessed in consideration of different disturbances. Moreover, as can be done to identify the specific environmental units of a building system, each activity can be assessed in relation to the others thanks to a compatibility-based approach. A similar choice can also be made for each element to define spatial and technological systems. For example, as introduced by Cavaglià et al. (1975), it is possible to suggest potential spatial organisations of activities (spatial units), using algorithmic calculations to define their mutual adaptation/disadaptation. Of course, specific requirements may be also interrelated with the prescriptive and morphological approach, like the minimal areas required for a single or double bedroom, or the minimal height of a spatial unit. Nevertheless, these mutual levels of disadaptation can be related to a list of requirements (e.g. temperature, humidity, air speed, privacy, sound level, individual space requirement, etc.) and expected levels of performance, by comparing the list of each activity. For example, washing hands, taking a bath/shower, going to the toilet, applying make-up and/or washing can be part of the same environmental unit. Several contemporary studies refer to the functional optimisation of the distribution of internal spaces, taking an algorithmic approach to spatial design (e.g. Medjdoub and Yannou 2001). Furthermore, similar approaches can be taken in the environmental early-spatial-organisation of environmental units, as in the case of studying sun and wind dynamics on a plot and further identifying the optimal locations for specific space units according to their related requirements (e.g. Chiesa and Grosso 2019a).

Although, it is essential to adopt a performance-driven approach to building design in the digital age—and not only the traditional prescriptive-morphological vision—this is related to the need for expression, and, in order to prevent a misunderstanding of outputs or poor usage of tools, it is essential for this performative vision to be based on specific knowledge (Carrara et al. 2014) and technical understanding of phenomena.

3.3 A Sample Application

The use of parametric design technologies has now become, at least in certain situations, part of professional practice. While, on the one hand, the use of integrated design is easier than in the past, also thanks to the progressive use of BIM platforms, often induced by regulatory requirements; on the other hand, the spread of tools for generative-computational design, or algorithmic procedural design, is still limited, at least in relation to design as a whole.

The potential of these computational/procedural tools allows the generation, assessment and/or optimisation of actions, with the introduction of scripts/models/algorithms—mechanisms and systems—related, for example, to specific forms, design choices and interventions. There are more and more examples of this, also thanks to progressive training of specific professional figures and in specific professional fields. Focusing on the software aspects and considering the potential expressed in the parametric field—as in the case of the Grasshopper environment for example—it is possible to overcome simple examples based on a few indicators, including additional checks, the results of which can be interconnected, ranging from the urban heat island estimation (e.g. using UWG), to advanced solar and lighting calculations, controlled natural ventilation flows, indoor/outdoor comfort indexes, dynamic energy simulations and system design, etc. Clearly, analysis is one of the steps of the approach, which will further include optimisation using algorithms, reiteration of the analyses and visualisation of results—see also Chap. 6 and Ref. (Chiesa 2013b). The enormous number of analyses that can be carried out on these platforms—see, for example, the toolbox of components compatible with grasshopper, not to mention the possibility of developing customised tools using compatible programming languages—and their mutual interrelations make up an almost unlimited panorama of cases which, on the one hand, are potential and, on the other hand, require the development of specific expertise to become actions.

In order to exemplify the potential of script-based parametric design, two examples of the shape of a surface are shown: the first aimed at modelling (1st digital age/sphere—see Sect. 3.3.1), and the second aimed at materialisation (2nd digital age/sphere—see Sect. 3.3.2). Although these scripts were developed several years ago—see in this sense (Casetta 2009; Chiesa and Casetta 2012)—they continue to be a valid example because, on the one hand, they relate to issues of constant interest to the designer (i.e. optimisation of a roofing surface on the basis of solar radiation; panelling of an initial surface, also with a double curvature, for mass-customisation) and, on the other hand, they can be replicated by the reader with the tools currently available, in order to investigate, also in direct contact, the potential of the world of parametric design without giving up its direct practical-professional implications. Both examples were created in the Rhinoceros® environment using Grasshopper™.

3.3.1 *Optimize a Shape by Using Specific Indicator—Conceptual Stage of Design (1st Digital Age)*

The paragraph focuses on the first of the two examples mentioned above, concerning the 1st digital age/sphere, in that it is based on a modelling project. In particular, it presents a script with the main aim of optimisation—understood as action (cybernetics), see (Ciribini 1968)—of the form of a roofing surface of a building on which photovoltaic panels are installed, in order to maximise—in terms of direction/action—the solar radiation that has an impact on these technical elements, considering the position and the surrounding area (as defined by Ciribini). This action is used as a model which can be applied to other technical elements—such as solar panels for the production of hot water, defining an appropriate domain/profile of use, e.g. schedule; or passive heating systems such as sunspaces—regardless of their surroundings, location and starting object. A description of the script that was developed to accomplish this goal is described below, using some simple examples of application. Table 3.1 shows the main elements of the algorithm developed.

For this example, the software Ecotect® was used to calculate the solar radiation impacting a surface. Ecotect® is a broadly used program in the field of environmental meta-design and is currently owned by Autodesk (Ecotect Analysis), however, the sale of new licenses was discontinued in 2015. This software, as well as others available on the market, allows the performance of solar analysis, sun and shadow studies, daylight and lighting, but also weather data visualisation, and simplified energy analysis of a building. Among others, it is possible to remember the following early-design tools: Heliodon v2 (see Beckers and Masset 2011), which is a "tool designed to control energy-related and visual aspects of natural lighting in urban and architectural projects", Dial+ (Estia 2017) an energy simulation tool for engineers and architects focusing on both lighting and cooling aspects with special regards to low-energy/natural solutions, and focusing on the Grasshopper/Rhinoceros environment, tools such as Diva and Alfa (Solemma 2016), or Ladybug tools (Ladybug, Honeybee, Butterfly, Dragonfly) released under the GPL-3.0 licence—see https://www.ladybug.tools/, viewed Jan 2019.

Table 3.1 Main elements of the example

Objective	Variables	Constraints	Software
Optimization of shapes in order to maximize the incident solar radiation on a photovoltaic roof	Several possibilities. E.g. roof generating points, maximum and minimum height, rotation axis (orientation), on-site location	Site location and weather data Site context (shadows and external conditions) Dimensional subdivision of photovoltaic panels	Rhinoceros®; Grasshopper® Geco^{cc} Ecotect Analysis® Galapagos

The analyses carried out using Ecotect are managed directly by the Rhinoceros + Grasshopper platform, making it possible, on one hand, to automatically process the calculation (regardless of the time taken by the processor for the long optimisation process), and on the other, to overcome the difficulties that may arise when a geometry must be imported/transferred to another environment, often requiring direct intervention on the geometries, something which is not contemplable if you want to be able not only to perform a single analysis but to implement a process of optimisation. In this specific case, the interoperability of the operation is managed by the plug-in for Grasshopper™ Geko^CC: the Ecotect software only needs to be installed on the computer and is managed in the background, both in terms of start-up and, using the LUA scripting language, for the management of I/O flows, settings and running. The calculation of radiation is carried out according to an processing rationale aimed at optimisation, integrating, in a need-performance perspective, the definition of the geometry and its assessment, simultaneously managing geometry generation, performance analysis, visualisation and feedback cycle.

The optimisation process is here based on the use of a devoted algorithm included in a plug-in of the Grasshopper™ sub-platform. Optimization algorithms (such as nesting—manufacturing sector) can be used thanks to contemporary computer performances. Evolutionary algorithms—e.g. genetic algorithms—(Fogel 1964; Fogel et al. 1966; Fogel 1999) helps to solve several architectural issues such as the optimization of shapes using specific indicators—see also (Weise 2009). The example presented below is based on genetic algorithms, in particular on the use of the Grasshopper's plug-in Galapagos (see also Rutten 2010). These algorithms are quite interesting because they allow for multi-criteria and multi-variable analysis. Of course in order to use an algorithm of this kind, long elaboration times are required, also because each iteration has to be solved completely, including the time used by the simulation software. Final output presents some uncertainty, as it does not guarantee a single final solution, working on a continuous flux that can be interrupted by using a limitation in the number of iterations or in time. This potential uncertainty has to be critically interpreted and solved by designers but represents a good starting point for early design steps also because intermediary solutions can be checked and the run-time process is transparent, helping the algorithm to overcome any obstacles. Focusing on Galapagos, after defining the number of variables (genes), the plug-in starts to populate a fitness landscape—a multidimensional spatial model in which the number of dimensions refers to variables and testing—using random genomes because the landscape geometry is unknown. Each solution is then assessed by the algorithm in order to define its suitability. According to these values, the algorithm searches for landscape peaks to start new iterations, focusing on better genomes rather than on the worse ones to find hypothetical peak in the fitting landscape. Figure 3.3 shows (a) a sample fitness landscape elaborated in Galapagos, (b) genome solutions, (c) the list of the genomes in an iteration step, and (d) a graph reporting the progressive evolution of testing values obtained through the iteration process. In this last graph, the points at the bottom of the curve correspond to a potential peak—not every solution defines a peak.

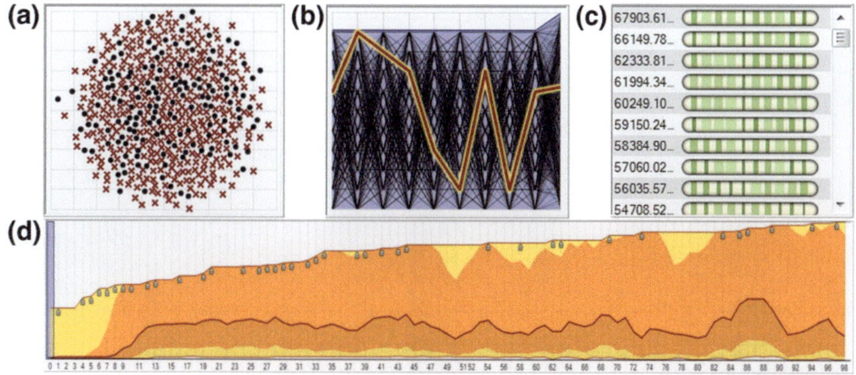

Fig. 3.3 Sample visualisations obtained by Galapagos concerning. **a** A sample landscape fitness; **b** genome solutions; **c** sample list of genomes used in an iteration step; **d** progressive evolution of the obtained testing values iteration by iteration, the points correspond to a new peak value

It is important to remember that this step can take several hours or even days to calculate.

Nevertheless, before using the genetic algorithm, the elaborated script envisages different steps. Firstly, it is necessary to (i) model the context and (ii) a generic shape for the selected building—a process that can be done in Rhinoceros by further connecting geometries in Grasshopper through geometry tools. After modelling the site, (iii) typical weather data and a corresponding file (*.wea) is implemented—different sources of a typical meteorological year may be found, such as the Energy-Plus weather database which is accessible on-line (https://energyplus.net/weather, viewed Jan 2019) including hourly-defined *.wea data, or the Meteonorm weather database which reports climate data with different extensions and time-step definitions (see also Meteotest 2016). At the same time, (iv) geographic coordinates and orientation are defined. Secondly, the script requires (v) the definition of design variables: fixed and movable primitives and/or geometries of the building, ranges of translation, dimensional parameters, deformation of the roof, multiple interrelations between variables, rotation axis and range (changes in orientation). For the proposed algorithm there are three typologies of variables: the boundary values (e.g. position, context); design constraints (e.g. laws, technical requirements, user needs); and design solutions (e.g. architectural choices, dimensions). It is essential to (vi) correctly evaluate ranges, relations and variable steps. In the end, it is possible to (vii) define a fitness function to evaluate the performances of selected geometries and maximize total incident solar radiation on specific surfaces (based on an analysis grid), which is directly calculated by the software Ecotect® (or similar). The fitness function is a critical point for generative algorithms, in fact, for multi-parameter optimizations the correct definition of evaluation indicators and weights is a non-banal issue. At this point the genetic algorithm becomes effective. Figure 3.4 illustrates the general process.

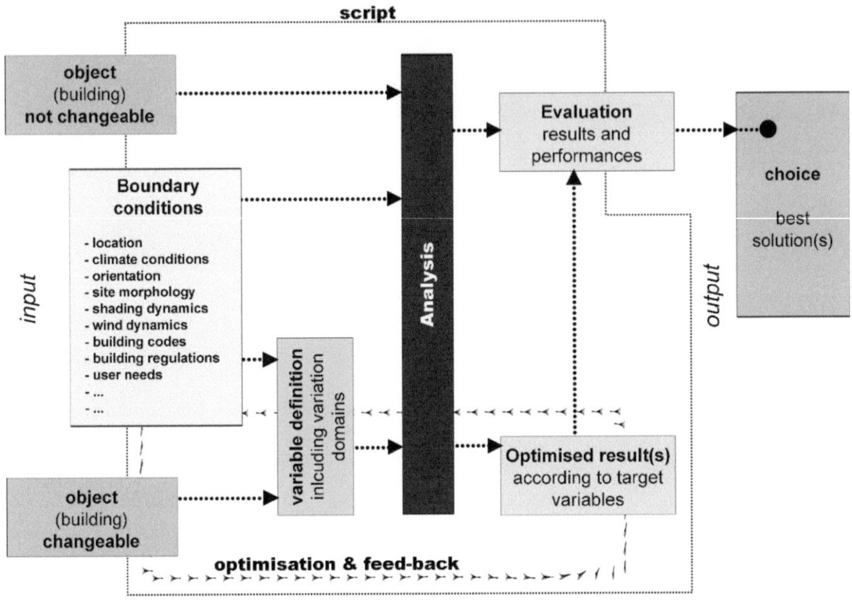

Fig. 3.4 The conception behind the defined algorithm, see boundaries between operator inputting, output critical evaluation and the scripting part

The script was applied to a hypothetical building with a rectangular plan, acting on the shape of the roof. For clarification purposes and to speed up the simulation process, the distribution and functional aspects of the building were reduced to the minimum habitable height only, while being aware that building design is more complex and that the procedural tools allow management of a greater number of variables (time and computing power permitting).

In order to make the action of the script easier to understand, several increasingly complex input scenarios were chosen. In particular, identifying three different configurations of the variables processed by the genetic algorithm (I, II, III) and two site/context configurations (a, b), with a total of six scenarios (I.a, I.b, II.a, II.b, III.a, III.b)—see also Fig. 3.5. All the scenarios were tested in the Torino Caselle site, using the relative weather file.

The three configurations of variables/genes are:

I. 3-variable configuration—(1) orientation of the building through rotation on the vertical central axis; (2)–(3) variation of the heights of the two main sides of the upper face;
II. 5-variable configuration—(1) orientation of the building through rotation on the vertical central axis; (2)–(5) variation of the height of the four upper vertices;
III. 10-variable configuration—(1) orientation of the building through rotation on the vertical central axis; (2)–(5) variation of the height of the four upper vertices; (6)–(9) variation of the height of the midpoints of the four sides of the upper face; (10) variation of the height of the midpoint of the upper face.

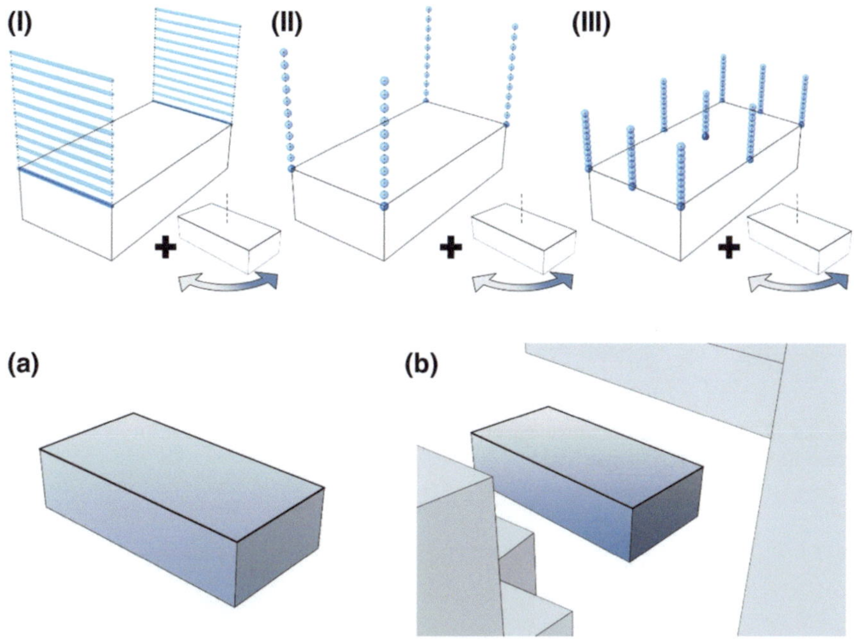

Fig. 3.5 Schematic representation of considered variables and the site configurations

The two configurations of the context are:

a. absence of obstructions, site free from shade restrictions (flat land);
b. sample urban environment with surrounding buildings.

Scenario I.a (S1)

Range and steps of variation of the three genes considered:

1. rotation on the vertical axis in the range ±90° with steps of 1°.

2.–3. variation of the heights in the range 0–600 cm with steps of 6 cm (0–100)—the height 0 corresponds to the minimum habitable height.

The number of processes calculated was 74. The solar panels considered are 160 * 80 cm in size for an uptake area of 1.28 m^2. The shape of the roof is automatically divided into panels of these dimensions. The number of panels, in this example, depends on the size of the upper face. Nevertheless, in addition to the total incident solar radiation on the entire surface, the value for area unit (m^2) is also given in the following examples.

Figure 3.6 shows some of the intermediate results obtained respectively (Fig. 3.4a, case S1.0) in the initial configuration—in other words with the genes positioned on 0, 0, 0 (Fig. 3.6b, case S1.1) after the first iteration—genes set to 16, 24, 96, where the last two values are in the range 0–100 (so 144 and 576 cm respectively) (Fig. 3.4c, case S1.5) after the fifth iteration—genes set to—38, 3, 95 (Fig. 3.6d,

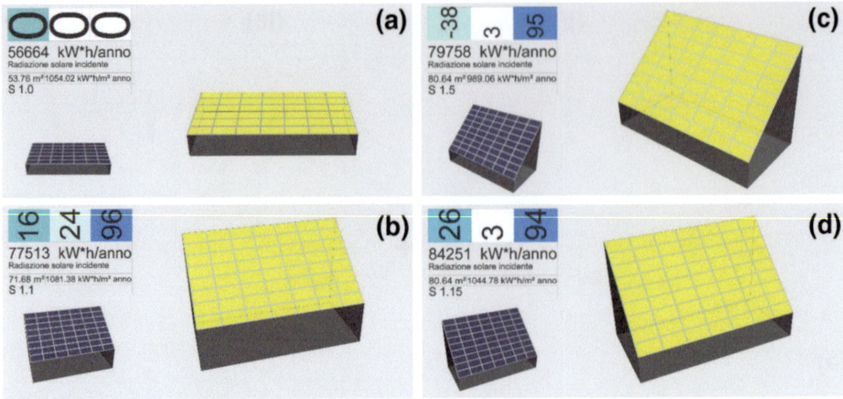

Fig. 3.6 Results of some iterations generated by Galapagos for the first scenario: **a** case S1.0; **b** case S1.1; **c** case S1.5; **d** case S1.15

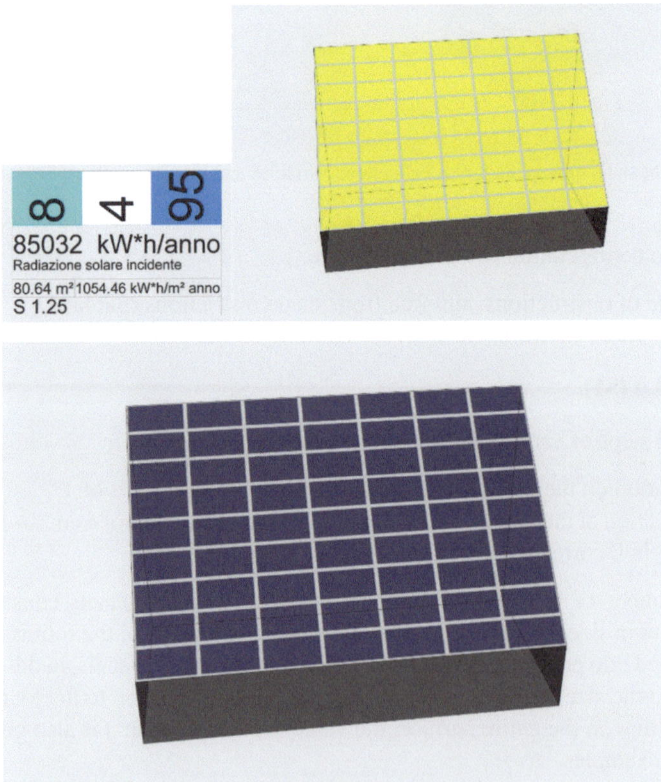

Fig. 3.7 Optimised shape reached at the 25th iteration

case s1.15) after the 15th iteration—genes 23; 3; 94. Figure 3.7 shows the optimised geometry. Considering the simplicity of scenario 1, it is possible to compare the result with that expected. As expected, the difference between the two heights allows the installation of the largest possible number of photovoltaic panels, considering the range of variation of the heights imposed (nine rows of seven panels). The optimal theoretical orientation of the photovoltaic panels in the absence of obstructions should be South, a value from which the peak found differs slightly. The angle of inclination is also equal to 47.52°, i.e. compatible with the angle closest to 35°—the maximum theoretical value for the latitude of the site without considering restrictions relating to the technical elements (e.g. panel dimensions)—suitable for locating the nine rows of panels. The optimal angle would be 46.02° considering that the inclined edge should measure 7.2 m (80 cm × nine rows). By increasing the number of processes, it is likely that the optimisation program will reduce this distance even more. As mentioned above, the interpretation of the results is important. Especially because, as we will see in the following cases, especially when the analysis is carried out in a built environment, optimal reference values are not always known.

Scenario II.a (S2)

Range and steps of variation of the five genes considered:

1. rotation on the vertical axis in the range ±60° with steps of 3°.
2.–5. variation of heights in the range 0–600 cm with steps of 60 cm (0–10)—the height 0 corresponds to the minimum habitable height.

The number of processes calculated was 54. The solar panels considered are the same as in the previous case.

The increase in variables involves an increase also in processing times, which is why the steps of variation have been altered in order to reduce the number of possible cases to be tested. Also in this case, some of the results obtained from subsequent processes are reported. Figure 3.8 shows (a) the initial case—S2.0—i.e. a flat roof; (b) the result of the 11th iteration, where the orientation differs by 3° from South, while the inclination of the pitch has increased until eight rows of panels can be accommodated, which in subsequent versions will become nine; (c) 25th iteration; (d) 29th iteration. Figure 3.9 shows the result of the last iteration calculated, S2.54. It is possible to show how the algorithm does not stop when it reaches a peak but continues its search—the results of the Figure do not deviate, from the value of the 29th iteration, other than in the case of potential decimal places. This comparison highlights how the definition of the parameters makes the various cases different, because the solver does not necessarily reproduce the same path as scenario 1 for example but is oriented towards the search for the most suitable genomes. In this sense, the choice of the designer can guide and speed up the process by intervening on the definition of variables and ranges of variation or suggestion.

Scenario III.a (S3)

Range and steps of variation of the 10 genes considered:

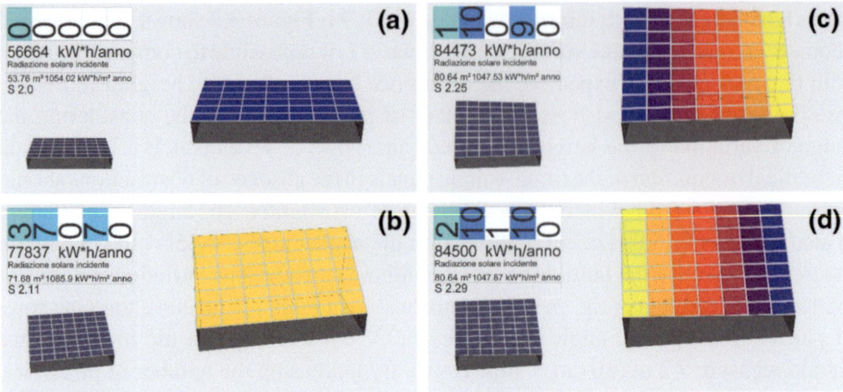

Fig. 3.8 Results reached by some of the calculated algorithm iterations: **a** initial case; **b** iteration 11th; **c** iteration 25th; **d** iteration 29th

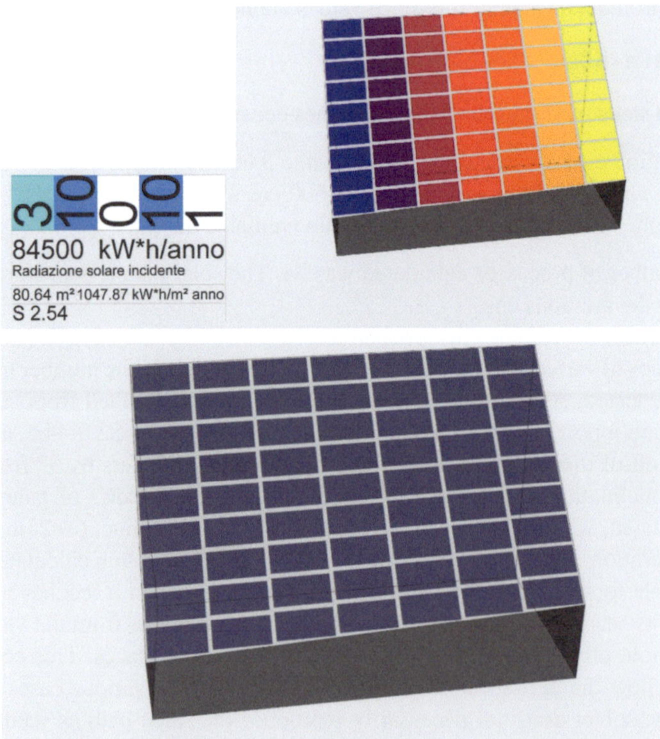

Fig. 3.9 Result of the last calculated iteration (54th). The figure underlines the fact that the solver has continued to modify genomes around a reached peck in the landscape fitness. The obtained result in term of annual kWh is, in fact, the same of Fig. 3.6d, such as the values of genes are very similar

1. rotation on the vertical axis in the range ±45° with steps of 1°.
2.–10. variation of heights in the range 0–300 cm with steps of 30 cm (0–10)—the height 0 corresponds to the minimum habitable height.

The number of iterations calculated was 98 + 80, with intervention at the 98th iteration, suggesting a direction to the solver—the first 98 iterations are classified in the figures as S3.1.*n*, the second 80 as S3.2.*n*. The size of the solar panels considered is the same as in the previous case, however the ranges of variation are no longer compatible—see the maximum height for each point of variation -, so a different result is expected.

Figure 3.10 shows, as in the previous cases, the results of some of the intermediate iterations calculated. In particular (a) shows the initial case S3.1.00; (b) the results of the first iteration, S3.1.01; (c) iteration S3.1.98, the last case of the first series; (d) the 32nd iteration of the second series, S3.2.32. Note how, also in this case, the value of the annual incident solar radiation increases in the various iterations. Lastly, Fig. 3.11 shows the result of the last iteration calculated. The third scenario further highlights how this process is able to optimise the incident solar radiation, this being the index chosen by the operator, considering the geometric variables imposed, but also the technical elements chosen (photovoltaic panels). The solution proposed also derives from the fact that the suggested roofing geometry is capable of housing a greater number of rows of panels than the original shape. Obviously, this result is an aid to the preliminary design processed by the architect, considering other aspects not included in this specific analysis, such as economic feasibility (e.g.: eliminating the panels that perform least well), technical feasibility, aesthetic taste, etc.

At this point, the same three simulations were carried out if the site was located in the urban area, where the surroundings were capable of generating shadows on the roof subject to analysis. These scenarios are particularly indicative for the value of algorithmic/procedural design because they are applied in contexts where it is

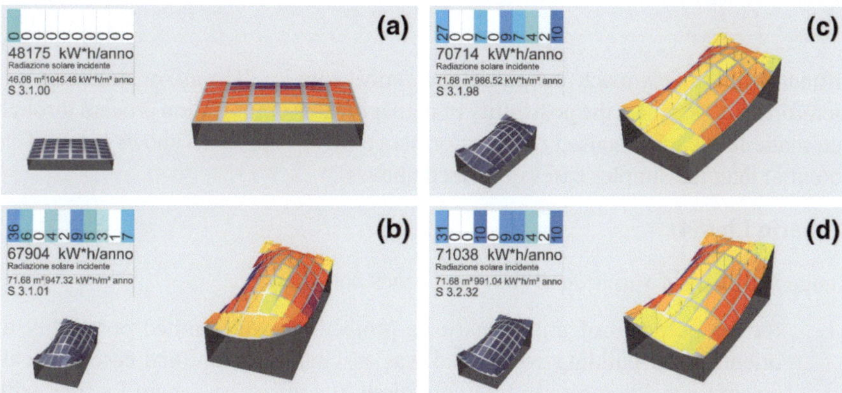

Fig. 3.10 Results of some of the iterations calculated by the algorithm, **a** starting case; **b** 1st iteration S3.1.01; **c** 98th iteration of the first series, S3.1.98; **d** 32nd iteration of the second series, S3.3.32

Fig. 3.11 Results of the last launched iteration

difficult to intuitively reach an optimal theoretical solution. The use of simulations is therefore necessary and the possibility of managing this optimisation process through automatic feedbacks focused on a performance-driven approach shows more of its potential than the simpler cases described above.

Scenario I.b (S4)

Range and steps of variation of the three genes considered:

1. for the purposes of this scenario, a project site with limited possibility of orienting the building considered was assumed. The optimal case was calculated considering only two conditions for this gene—main façades with East-West exposure (0) and North-South exposure (1), the entire surroundings are included in the rotation—the orientation is considered as restricted by the site and the two cases deal with two potential situations;

2.–3. variation of heights in the range 0–300 cm with steps of 30 cm (0–10)—the height 0 corresponds to the minimum habitable height.

The size of the photovoltaic panels considered is the same as in the previous scenarios. Figure 3.12 summarises the results obtained for some of the iterations

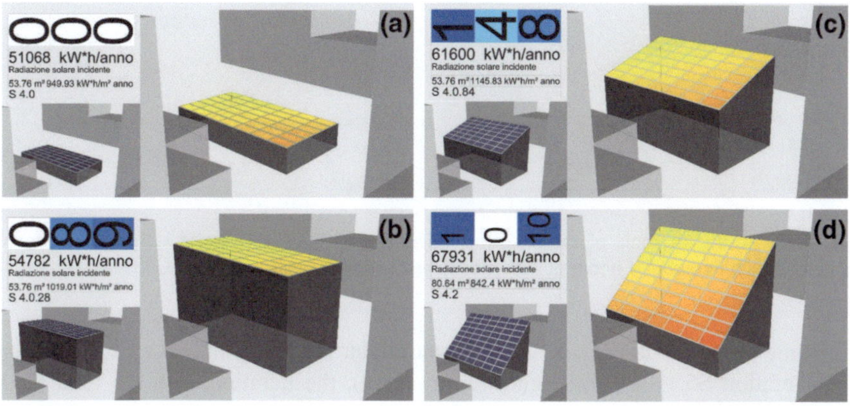

Fig. 3.12 Results of some of the iterations calculated by the algorithm. **a** Starting case, orientation 0; **b** 28th iteration, orientation 0; **c** one of the first iterations with orientation 1; **d** final simulated iteration with orientation 1

obtained. In particular, in (a), the initial case is shown, with East-West exposure of the façades (presumably not optimal); in (b), the 28th iteration, which also coincides with one of the radiation peaks found for exposure 0; in (c), one of the first iterations with the façades facing North and South (axis of the East-West building); in (d), the final peak found with this orientation. If the two main facades are exposed to East and West, the solution obtained by the algorithm, considering the need to maximise annual radiation, is a semi-flat roof, sufficiently high to reduce the incidence of shadows cast by the surroundings. If, on the other hand, the main façade faces south—the smaller facades are restricted to the heights of the sides of the larger ones—the roof with the greatest inclination towards this cardinal point will maximise the annual incident solar radiation, as the maximum height set is three metres, corresponding to an inclination angle of about 31°.

Scenario II.b (S5)

Range and steps of variation of the four genes considered:

0. for the purposes of this scenario, no rotations of the building were assumed and the main façade was oriented in a southerly direction;

1.–3. variation of heights in the range 0–600 cm with steps of 60 cm (0–10)—the height 0 corresponds to the minimum habitable height.

In this scenario, 43 iterations were calculated. The size of the photovoltaic panels considered is the same as in the previous scenarios. Figure 3.13 illustrates the results

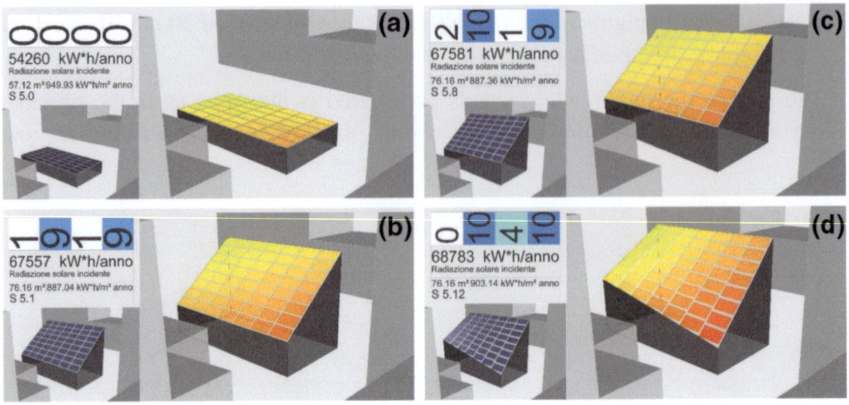

Fig. 3.13 Results of some of the iterations calculated by the algorithm, **a** starting case; **b** 1st iteration; **c** 8th iteration; **d** iteration with the highest recorded peak

for some of the iterations calculated, in particular (a), the initial case S5.0; (b), the first iteration S5.1; (c), the eighth iteration; (d), the iteration with the final peak identified. This scenario, allowing the independent movement of the single sides of the roof, orients the surface not only towards the south, as in the case of S4, but also towards the side on which the surrounding area has less influence.

Scenario III.b (S6)

Range and steps of variation of the nine genes considered:

0. for the purposes of this scenario, no rotations of the building were assumed and the main façade was oriented in a southerly direction;

1.–9. variation of heights in the range 0–300 cm with steps of 30 cm (0–10)—the height 0 corresponds to the minimum habitable height.

In this scenario, 83 iterations were calculated. The size of the photovoltaic panels considered is the same as in the previous scenarios. Considering the possibility of modifying up to nine points independently on the roof, the expected shape of this scenario, as in the case of the S3, is hard to imagine. Figure 3.14 shows this by reporting (a), the initial case; (b), the 1st iteration; (c), the 15th iteration; (d), the 66th iteration.

3.3.2 *From Shapes to Tessellation for Mass Customisation (2nd Digital Age)*

This point reports the second of the scripts developed to exemplify the potential of algorithmic/procedural design techniques combined with the performance-driven design approach.

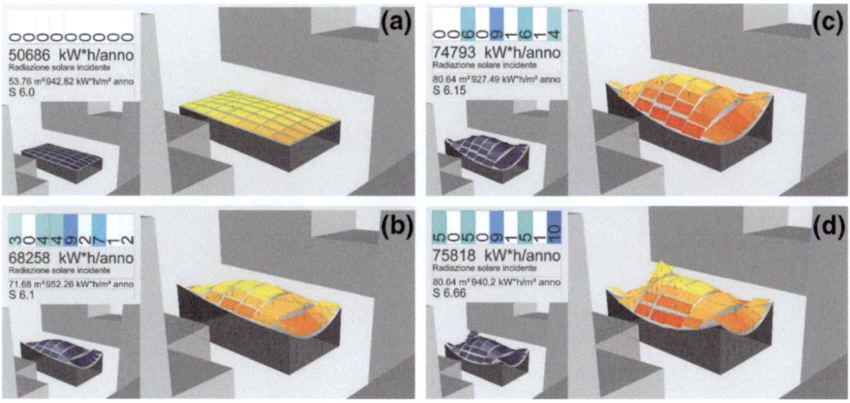

Fig. 3.14 Results of some of the iterations calculated by the algorithm, **a** starting case; **b** 1st iteration; **c** 15th iteration; **d** iteration with the highest recorded peak (66th)

Mass customization of different serial elements is a specific application for algorithmic design. In fact, the parameterization of processes allows for the management of a large number of unique elements produced in an industrialized methodology. With new software platforms it is possible to generate, analyze and manage complex geometries. This specific example illustrates an algorithmic process applied to a surface in order to obtain a panel system that could be produced with numerically-controlled machines. There are several examples which are able to reduce complex surfaces and shapes into panels; very complex solutions can be found in (Eigensatz et al. 2010). Especially for double-bend surfaces the complexity of the process requires an analysis of the deformation of panels.

In this case, the example refers to the design of a model to produce casing elements (panels) that can be adapted to cover any surface derived from predefined generating curves or directly from a predefined freeform surface. Each element is made up of five quadrilateral parts: a central panel, derived from the panelling of the surface to be reproduced, and four lateral flaps arranged on the sides of the central face, the primary function of which is to provide support for the fastening of the various main elements. The height of these flaps is a parameter that can be entered in the script and, assuming appropriate measurements, these surfaces can also assume a structural function. Each flap is designed in relation to its production, envisaged in laser cutter, to be bent at an angle on the edge of the main panel and welded to the side flap of the adjacent panel. The script generates it so that it can perfectly overlap the one next to it. The entire element—see also Fig. 3.15—is designed to adapt to the conformation of the original surface, and the size of the main panels is arranged to suit an average size (length and width) established during the definition of the input by the designer to adapt to production requirements. Moreover, the different surfaces are assessed with respect to the flexion envisaged by the panel, to check whether the material they are made can withstand the stress. If it cannot, the algorithm indicates a folding line for the individual faces, avoiding excess tension on the material and allowing the final

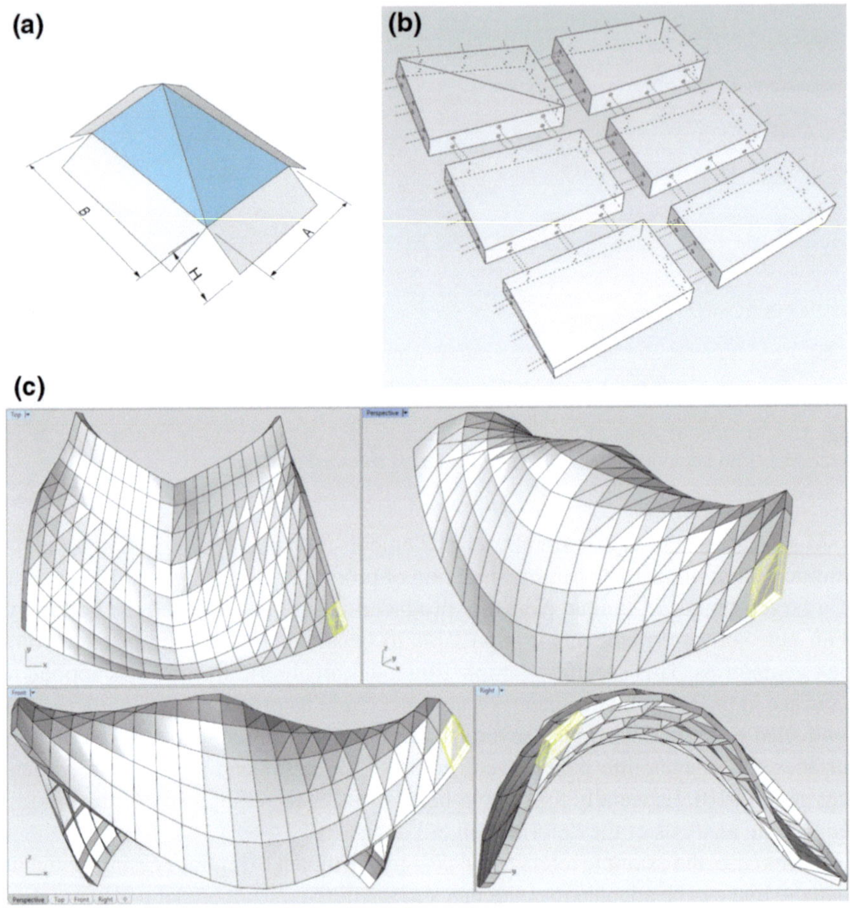

Fig. 3.15 **a** Sample visualisation of one of the panel produced by the developed script; **b** schematic view of panels' connections; **c** sample applications on randomly generated surface

construction of the structure. This fold will be highlighted on the sheet also during the production phase to facilitate the assembly of the various surfaces. Moreover, again in the laser cutting phase, each vertex of the flaps is numbered following an order generated automatically by the script indicating the vertices and adjacent flaps to which it will be connected. Table 3.2 illustrates the basic elements of the algorithm.

The entire process could be divided into three steps: (i) tessellation; (ii) two-dimensional representation of geometries; and (iii) fabrication.

The tessellation process used in the developed algorithm could be organized in sub-steps. Firstly, the original surface is defined; the entire process is based on a continuous shape that is generated in an independent way before the specific scripting process. Secondly, the script generates a tri-dimensional tessellation of the original surface based on the medium dimension of the panels. These geometries constitute

Table 3.2 Main elements of the example

Objective	Materials	Constraints	Software
Post-rationalization of complex surfaces in order to produce building-envelope elements which are able to fit every shape and size	Planar metal/plastic/glass surfaces cut by CNC cutter or laser cutting machines	Dimensional subdivision of planes in order to fit sheets (of metal, plastic, …) Deformation of panels in relation to the original shape and physically Chosen material (especially for double-bend surfaces)	Rhinoceros®; Grasshopper®; connected with mass customization machines

the tiles. The dimensions of flaps or related structures are implemented as parameters as shown in Fig. 3.16, which reports the developed script. Moreover, every tile surface is analysed for planarity and deformation, such as was mentioned before. The bending values are evaluated by an imputed limit, which changes according to the material, in order to subdivide non-suitable surfaces. Finally, every surface and relative flaps are deployed over a fixed plan, starting from the same point. This process is particularly important because it relates the generated panelization to a representation that constitutes the input of the mass-production process. The two-dimensional representation of geometries constitutes, in fact, a second step of (ii). Firstly, since every tile is deployed and planar but panels are overlapped, geometries

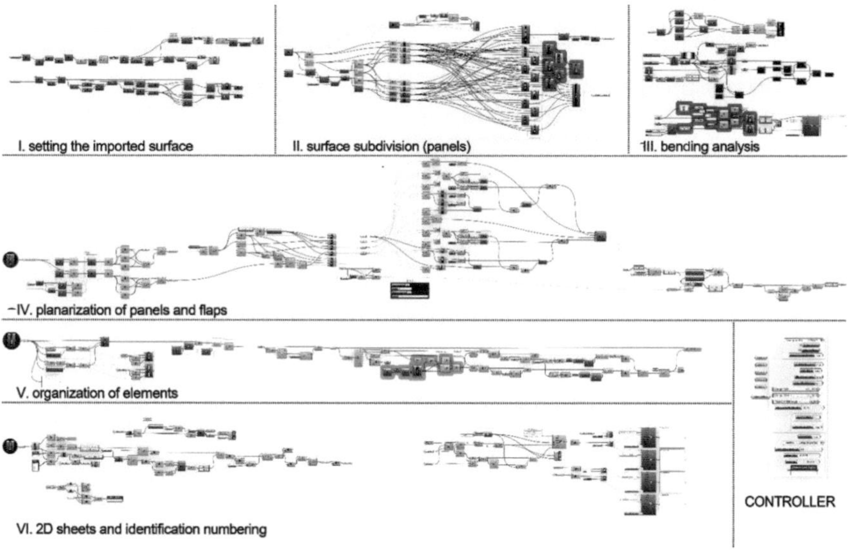

Fig. 3.16 Flow chart of the elaborated algorithm

are disposed in a specific order, according to final use. Tiles could be organized in a progressive way based on the starting position on the original shape; in this case mutual positions constitute a parameter. Panels can also be organised to reduce empty space between geometries in the x and y axes. A third solution involves organizing tiles in optimized solutions for mass production: nesting algorithms are necessary to reduce by-products from laser cutting techniques. Secondly, organized surfaces are tagged with specific codes, as mentioned before, in order to be identified during subsequent stages (e.g. assembling and maintenance).

At this point the CAD file with the organized surfaces is transformed into a readable input for a laser cutter machine in order to fabricate the panels. In the end, cut components are assembled and tested.

Figure 3.17 illustrates a hypothetical application of this methodology.

The definition of this specific algorithm has been developed to make it applicable to any surface. For descriptive purposes, four script application examples are given below.

The first (I) relates to a single-curved surface, generated by two equal profiles of curved section lying on parallel and aligned planes. Figure 3.18a shows the generating curves, and the three-dimensional model rendered by the algorithm—the colours represent the stress to which the panel is subject; Fig. 3.18b shows the cutting scheme of the components of the casing designed. Due to the simplicity of the surface, the panels are made up of squares.

The second (II) refers to a much more complex double curve surface, generated by non-planar curved section profiles with arbitrary orientations. Figure 3.19a shows the generating curves and the three-dimensional model of the panels including the results of the stress assessment. Figure 3.19b shows the cutting panel produced by the script. In this case, the shapes and dimensions of the panels are extremely variable, in order to cover the complex starting curve.

The third example (III) is related to a surface with a double curve, generated a priori. The construction lines are derived from the input surface. Figure 3.20a shows the construction lines derived from the original surface and the 3D model of the panels generated by the script. Figure 3.20b also shows the planar panels for their laser cutter production.

The fourth example (IV) is based on a surface generated a priori with a double curve. Figure 3.21 shows (a) the construction lines derived from the original surface

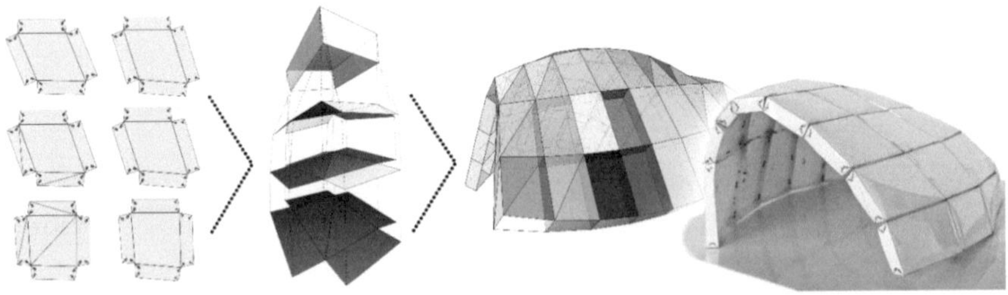

Fig. 3.17 Example: from 3D surface to 3D modelling and physical model/object

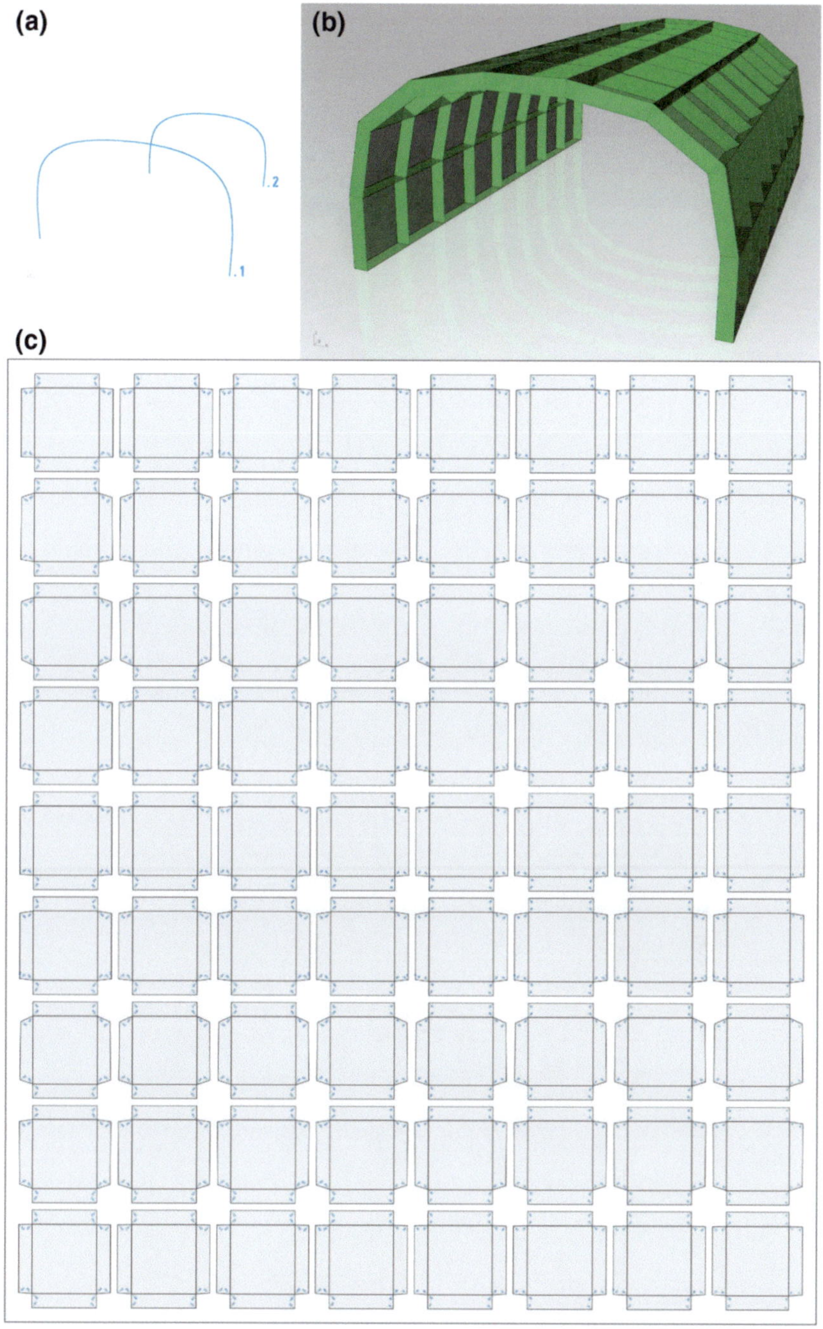

Fig. 3.18 Sample I, single-curvature surface. **a** Generating curves and **b** 3D model elaborated by the algorithm; **c** script-organised disposition on the cutting panel of all elements for their production

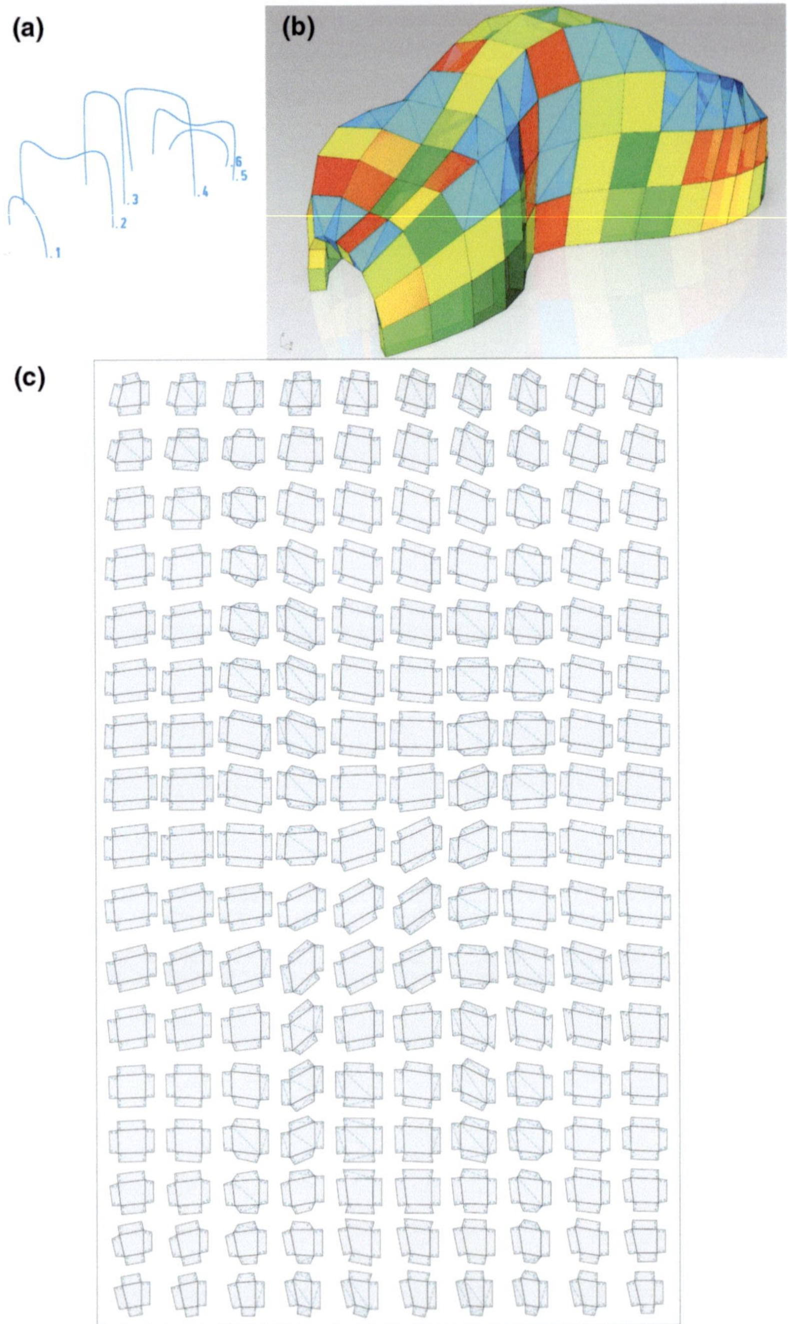

Fig. 3.19 Sample II, double-curved surface. **a** Generating curves and **b** 3D model elaborated by the algorithm; **c** script-defined disposition on the cutting panel of all elements for their production

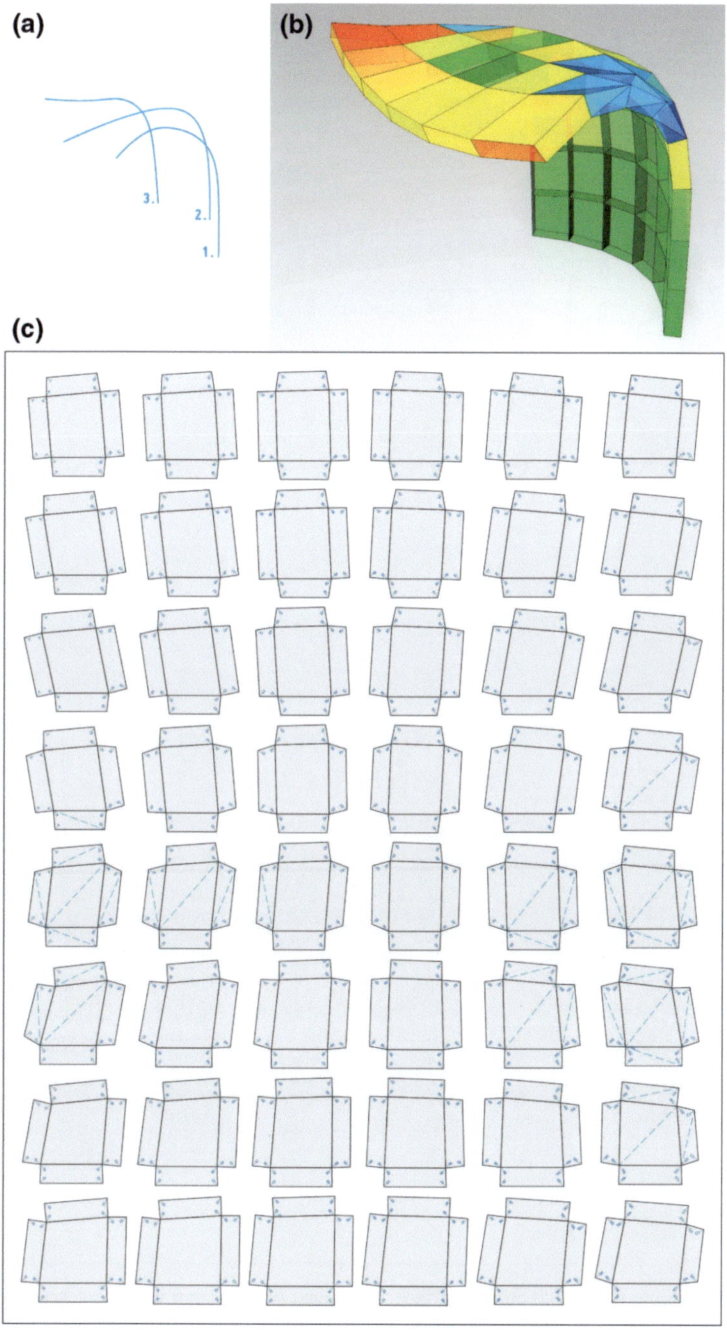

Fig. 3.20 Sample III, double-curved surface. **a** Constructive curves generated from the original surface and **b** 3D model elaborated by the algorithm; **c** script-defined disposition on the cutting panel of all elements for their production

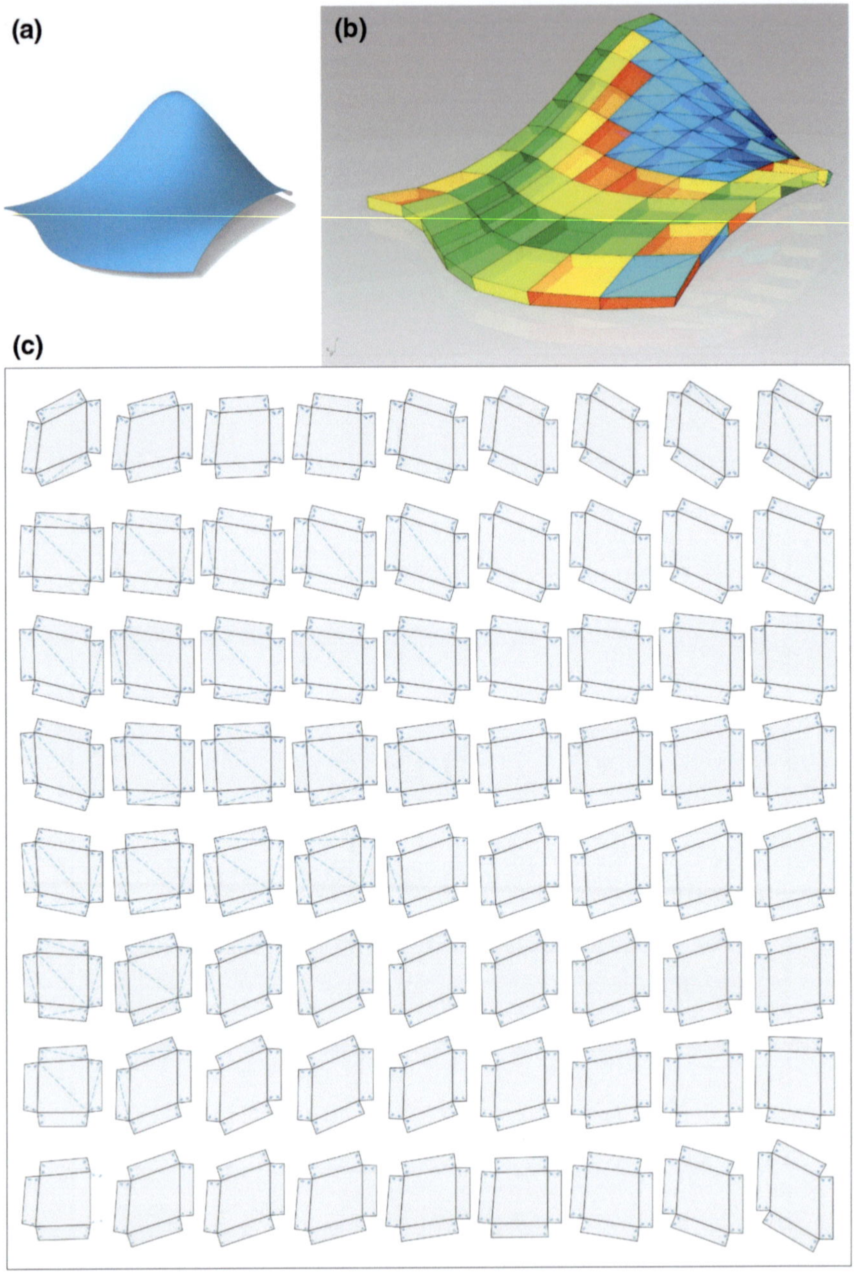

Fig. 3.21 Sample IV, double-curved surface. **a** Original surface from which constructive curves were generated and **b** 3D model elaborated by the algorithm; **c** script-defined disposition on the cutting panel of all elements for their production

and the three-dimensional rendering of the panels, and (b) illustrates the arrangement of the panels on the cutting sheet.

As shown by these examples, the script is a very versatile tool and, with a high initial development, a suitable approach for application to a wide variety of cases. Obviously, the designer's work is not limited to development, but also to the analysis of the results aimed at aligning that produced by the algorithm with project expectations.

3.4 Discussion and Conclusions

That discussed in this chapter highlights how it is possible, also thanks to the current tools and increasing computing power available, to orient the architectural design method specific of technological and environmental vision to design—based on a performance-driven ("esigenziale prestazionale") approach—focused on a vision "by programs" of the design itself. This vision, which refers to that introduced in Italy in the '60s and '70s in the discipline of technological architecture (e.g. Cavaglià et al. 1975; Ciribini 1968), finds a new application in the contemporary world. Previous theories, the applications of which were limited due to a lack of adequate tools, would be feasible today, although they would be achievable mainly because current applications and potentials are set within in a new horizon, in which the frontiers of the different types of knowledge involved are no longer those of archetypal visions. The examples shown must be read and understood in this sense, where it is demonstrated that it is possible to use the computational generative process to optimise forms based on specific constraints in a process that links modelling to materialisation.

The need-performance driven method presented here can be perfectly integrated with the new way of creating design—see also Chap. 5. In this method, the action of design and its work in progress (WIP) are explicitly linked: (i) to the parties involved—whose involvement also takes place through iterative phases, as long as that input and output can be translated between the different languages involved, (ii) to the aspects specific to the design—i.e.: classes of needs, including those aspects that other areas related to the discipline define as aesthetics, functionalities (e.g. distributive, structural, plant engineering, etc.), sustainability, etc. (iii) the design phases—from early design ("metaprogetto" e.g. building programming—see also Magnaghi 1967) through to construction aspects, and the subsequent design of maintenance or disposal operations. These explicit links with design are supported by the framework of skills, understanding (technical/theoretical), culture, and networks linked to the necessary connective layers related to interfaces and interoperability. These two connective aspects are essential to be able to act within the field of parametric design, allowing the exchange of information and the application of various expert skills.

Obviously, this method can be adapted to the most diverse processes and areas of application. If we continue looking at building, for example, it is possible to connect geometrical tools, equipped with a visual interface for design—e.g. CAD—with other information/decisional/performative areas of design, related to technological, environmental and energy aspects for example. The parametric study of how the alteration of certain parameters based on choices (e.g. technical, aesthetic, …) changes the impact on the context and on the building/subject of analysis opens up the way to values not only of a practical-professional nature, but in terms of education too. If this process is explicit and not implicit, the results will lead to a progressive increase in accumulated knowledge. Moreover, the various design aspects are interconnected and, in the world of parametric design, the single constraint related to an a priori choice of a certain value to be associated with a technical element can be overcome by a holistic vision capable of including the interactions of that value with the context and the different systems connected to it. Supporting the passage from a tree vision to a pattern approach, in which the exchange of information and relations between the parties concerned becomes an open but concatenated structure (David and Oppio 2016; Claudel et al. 2016; Alexander 1965).

Didactically, the methodological vision of need-performance-driven design can be pursued even in the absence of a platform capable of making the various interfaces interact and allowing the interoperability of information, replacing this information processing tool with the direct action of the designer, which is always necessary—see also Sect. 1.3—as demonstrated in the didactic methodologies of the technological field (e.g. Chiesa et al. 2018b; Chiesa and Grosso 2017b). However, it is in the potential of the parametric computational tool that this methodological horizon can be extensively applied (e.g. Chiesa and Grosso 2019a), both didactically and professionally, making it possible to address individual problems or more complex areas of interrelation between the parties. This vision may include several specific aspects. It is possible to link design with energy-dynamic simulators (e.g. Chiesa et al. 2018a, 2019) or, switching from design to reality, the building with simulators aimed at optimising its operation.

The possibility of defining the method, the action, rather than the single element to be repeated identically every time, is well linked to design, which, by its very nature, is specific and must, or at least should, adapt to different contexts and environments, reacting to mutual interactions. In this sense, the possibilities opened up by IT in the field of computational design—which we can expect to include and adapt to data deriving directly from the environment by integrating into a complete ICT vision—open up new creative road that allow innovative interactions between science and art, and not their limitation as it might seem in some cases. The methodological process is integrated, allowing the pre-rationalisation of systems, forms and their decomposition into sub-components, bringing processes closer together and avoiding retrofit actions. Think, for example, of the development of complex surfaces, of the necessary actions aimed at rationalising them, whether for construction, economic, physical-technical, environmental or other purposes—while being aware that rationality is never absolute, as design must be understood as a development, which cannot be free of doubt and of the assessment/design decision, which doesn't claim to be absolute and unconditional. This rationalisation, however, linking conceptual knowledge to know-how, can reduce, at least theoretically, the deviation that often occurs between design and object.

Moreover, the parametric way of thinking can easily become a metaphor or an analogy of adaptation, where the concept of parameterisation/optimisation of the design geometry transposes itself into movement of parts of the building/object, optimising itself in real-time with the real context. Examples of this are the work on kinetic/responsive façade systems, e.g. the Aegis hyposurface, Mark Goulthorpe and Decoi Group, 1999/2000 Birmingam Hippodrome Theatre, UK; the Kiefer technic showroom, Ernst Giselbrecht and Partners, 2007, Styria, Austria; the Digital Water Pavilion, Carlo Ratti Associati, Expo 2008, Zaragoza, Spain. Also in this sense it is possible to link—see Chap. 2, but also the EU registered design "Bio_Logic Skin" (Chiesa et al. 2018c; Olivero 2017)—the modelled geometry to the real object by creating a series of analogies, relationships, optimisations.

References

Abbagnano N (1961) Dizionario di filosofia. UTET, Torino

Alexander C (1964) Notes on the synthesis of form. Harvard University Press, Cambridge

Alexander C (1965) A city is not a tree. Archit Forum 122(1):58–62

Allen E (2005) How buildings work. The natural order of architecture, 3rd edn. Oxford University Press, Oxford

Atkinson RD (2004) The past and future of America's economy: Long waves of innovation that power cycles of growth. Edward Elgar, Cheltenham, UK

Beckers B, Masset L (2011) Heliodon version 2.7-3. Heliodon 2 documentation. Accessible at http://heliodon.net/heliodon/index.html. Last view Jan 2019

Bocco A, Cavaglià G (2008) Cultura tecnologica dell'architettura. Pensieri e parole, prima dei disegni. Carocci, Roma

Bosia D (ed) (2013) L'opera di Giuseppe Ciribini. FrancoAngeli, Milano

Carrara G, Fioravanti A, Loffreda G, Trento A (2014) Conoscere collaborare progettare. Gangemi, Roma

Casetta E (2009) Parametri e architettura temporanea: progettare il prototipo. Master Degree Thesis, MArch in Architecture, Politecnico di Torino, Academic Year 2010/11, Supervisor: R. Pagani, co-Supervisor: G. Chiesa

Cavaglià G, Ceragioli G, Foto M, Maggi PN, Matteoli L, Ossola F (1975) Industrializzazione per programmi. Strumenti e procedure per la definizione dei sistemi di edilizia abitativa. Studi e Ricerche RDB, Piacenza

Celento D (2007) Innovate or Perish. New technologies and architecture's futures. Harvard Des Mag 27:1–9

Chiesa G (2013b) La città digitale, dai sensori ai modelli: Piattaforme interconnesse per la città del futuro. In: Di Giulio R et al (eds) Strategie di riqualificazione urbana: Rigenerazione e valorizzazione dell'edilizia sociale ad alta densità abitativa del secondo Novecento. Quodlibet, Macerata, pp 110–117

Chiesa G (2016) Model, digital technologies and datization. Toward an explicit design practice. In: Pagani R, Chiesa G (eds) *Urban data. Tools and methods towards the algorithmic city.* FrancoAngeli, Milano, pp 48–81

Chiesa G, Casetta E (2012) Parametric CAD modelling and scripting for architectural technology (unpublished paper)

Chiesa G, Grosso M (2017a) An environmental and technological approach to architectural programming for school facilities. In: Sayigh A (ed) Mediterranean green buildings & renewable energy. Springer, Amsterdam, pp 701–715

Chiesa G, Grosso M (2017b) Environmental and technological design: a didactical experience towards a sustainable design approach. In: Gambardella C (ed) World heritage and disaster. Knowledge, culture and representation, Le Vie dei Mercanti_XV International Forum. La scuola di Pitagora Editrice, Napoli, pp 944–953

Chiesa G, Grosso M (2019a) A parametric tool for assessing optimal location of buildings according to environmental criteria. In: Sayigh A (ed) Sustainable building for a cleaner environment. Springer, Cham, pp 115–130

Chiesa G, Grosso M (2019b) Meta-design approach to environmental building programming for passive cooling of buildings. In: Sayigh A (ed) Sustainable building for a cleaner environment. Springer, Cham, pp 285–296

Chiesa G, Grosso M, Acquaviva A, Makhlouf B, Tumiatti A (2018a) Insulation, building mass and airflows—provisional and multi-variable analysis. SMC-Sustain Mediter Constr 8:36–40

Chiesa G, Grosso M, Ahmadi M, Bo M, Murano G, Nigra M, Primo E (2018b) Design of indoor climate control passive systems in buildings: experiences for a PhD course. In: Gambardella C (ed) World heritage and knowledge. Representation, restoration, redesign, resilience, Le Vie dei Mercanti_XVI International Forum. Gangemi International, Roma, pp 229–237

Chiesa G, Olivero D, Griffa C (2018c) Bio_Logic Skin. Disegno o Modello comunitario registrato, No. 005510716–0001, EU IPO, European Union Intellectual Property Office

Chiesa G, Acquaviva A, Grosso M, Bottaccioli L, Floridia M, Pristeri E, Sanna EM (2019) Parametric optimization of window-to-wall ratio for passive buildings adopting a scripting methodology to dynamic-energy simulation. Sustainability 11:30. https://doi.org/10.3390/su11113078

Ciribini G (1968) Brevi noti di metodologia della progettazione architettonica. Edizioni quaderni di studio, Politecnico di Torino, Torino

Ciribini G (1970) I componenti nel "performance design". Politecnico di Torino, Torino

Claudel M, Pedrazzo MM, Suraci N (2016) Raster to vector: towards a live associative model. In: Pagani R, Chiesa G (eds) Urban Data. FrancoAngeli, Milano, pp 106–117

Conole G, Wills S (2013) Representing learning designs—making design explicit and shareable. Educ Media Int 50(1):24–38

David A, Oppio A (2016) Combining pattern theory with spatial multicriteria analysis for urban planning. The case of a neighborhood renewal in Turin (Italy). In: Pagani R, Chiesa G (eds) Urban data. FrancoAngeli, Milano, pp 121–158

Davies N, Jokiniemi E (2008) Dictionary of architecture and building construction. Elsevier, Amsterdam

Droege P (2006) The renewable city: a comprehensive guide to an urban revolution. Wiley, Chichester

Eigensatz M, Deuss M, Schiftner A et al (2010) Case studies in cost-optimized paneling of architectural freeform surfaces. In: Ceccato C, Hesselgren L, Pauly M, Pottmann H, Wallner J (eds) (2010) Advances in architectural geometry 2010. Springer, Vienna, pp 49–72

ESTIA (2017) Dial + Version 2.5 Manuel d'utilisation. Visible at: http://docs.wixstatic.com/ugd/4e84bd_96cc28d2498749d5a963c6fbe29837c1.pdf. Last view Jan 2019

Ferraris M (2003) Ontologia e oggetti sociali. In Floridi L (ed) Linee di Ricerca. SWIF, p 269–309, viewed Jan 2014 <www.swif.it/biblioteca/lr>

Fogel LJ (1964) On the organization of intellect. Ph.D. thesis, UCLA University of California, Los Angeles, California, USA

Fogel LJ (1999) Intelligence through simulated evolution: forty years of evolutionary programming. Wiley-Interscience, New York

Fogel LJ, Owens AJ, Walsh MJ (1966) Artificial intelligence through simulated evolution. Wiley, New York

Freeman C, Perez C (1988) Structural crises of adjustment: business cycles and investment behaviour. In: Dosi G et al (eds) Technical change and economic theory. Printer, London, pp 38–66

Grosso M (2005) Valutazione dei caratteri energetici ambientali nel metaprogetto. In: Grosso M, Peretti G, Piardi S, Scudo G (eds) Progettazione ecocompatibile dell'architettura. Esselibri, Napoli, pp 307–336
https://www.ladybug.tools/. Last view Jan 2019
https://energyplus.net/weather. Last view Jan 2019
Kalay YE (2006) The impact of information technology on design methods, products and practices. Des Stud 27(3):357–380
Kant I (1787) Kritik der reinen Vernunft (Italian translation (2000) Critical della ragion pura. Laterza, Roma-Bari)
Magnaghi A (ed) (1967) Contenuto e funzione del metaprogetto in architettura. CNR, Politecnico di Torino, AIRE, Milano
Mansfield M (1983) Long waves and technological innovation. Am Econ Rev 73(2):141–145
Matteoli L (2013) Prof. Ing. Giuseppe Ciribini. In: Bosia D (ed) L'opera di Giuseppe Ciribini. FrancoAngeli, Milano, pp 147–150
Medjdoub B, Yannou B (2001) Dynamic space ordering at topological level in space planning. Artif Intell Eng 15:47–60
Meteotest (2016) Meteonorm global meteorological database v. 7.1—Handbook Part I: Software. Meteotest, Bern, p 32
Olivero D (2017) Maker in architettura: esperimenti di fabbricazione di una Responsive Surface. Master Degree Thesis, MArch in Architecture for the sustainability design, Politecnico di Torino, Academic Year 2017/18, Supervisor: G. Chiesa, co-Supervisor: C. Griffa
Oxman R (2006b) Editorial. Special issue of design studies on digital design. Design Studies 27(3):225–227
Piano R (2002) La responsabilità dell'architetto. Passigli Editori, Firenze
Reid E (1988) Understanding buildings. A multidisciplinary approach. Longman Scientific and Technical, London
Reigeluth CM, Carr-Chellman AA (2009) Instructional-design theories and models. Building a common knowledge base, vol III. Routledge, New York
Rothenberg J (1989) The nature of modelling. In: Widman LE, Loparo KA, Nielson NR (eds) Artificial intelligence, simulation & modelling. Wiley, New York, pp 75–92
Rutten D (2010) Evolutionary principles applied to problem solving. Accessible at: https://www.grasshopper3d.com/profiles/blogs/evolutionary-principles. Last view Jan 2019
Sennet R (2008) The craftsman. Yale University Press, New Haven
Simon HA (1981) The sciences of the artificial. MIT Press, Cambridge
Šmihula D (2009) The waves of the technological innovations of the modern age and the present crisis as the end of the wave of the informational technological revolution. Studia Politica Slovaca 1/2009:32–47. https://papers.ssrn.com/sol3/papers.cfm?abstract_id=2353600. Last view Apr 2019
Solemma (2016) see https://www.solemma.com/index.html. Last view Jan 2019
Torricelli MC, Del Nord R, Felli P (2001) Materiali e tecnologie dell'architettura. Laterza, Roma-Bari
Tosoni P (2008) Il gioco paziente. Biagio Gardena e la teoria dei modelli per la progettazione, 2nd edn. Celid, Torino
UNI (1978) Edilizia. Terminologia per requisiti e prestazioni. Verifiche di conformità relative ad elementi, UNI 7867-3:1978. Ente Nazionale Italiano di Unificazione, Milano
UNI (1978) Edilizia. Terminologia per requisiti e prestazioni. Specificazione di prestazione, qualità e affidabilità, UNI 7867-2:1978. Ente Nazionale Italiano di Unificazione, Milano
UNI (1978) Edilizia. Terminologia per requisiti e prestazioni. Nozioni di requisito e di prestazione, UNI 7867-1:1978. Ente Nazionale Italiano di Unificazione, Milano
UNI (1979) Edilizia. Terminologia per requisiti e prestazioni. Qualità ambientale e tecnologica nel processo edilizio, UNI 7867-4:1979. Ente Nazionale Italiano di Unificazione, Milano
UNI (1981) Edilizia. Esigenze dell'utenza finale. Classificazione, UNI 8289:1981. Ente Nazionale Italiano di Unificazione, Milano

UNI (1983) Edilizia residenziale. Sistema tecnologico. Analisi dei requisiti, UNI 8290-2:1983. Ente Nazionale Italiano di Unificazione, Milano

UNI (1999) Edilizia - Terminologia riferita all'utenza, alle prestazioni, al processo edilizio e alla qualità edilizia, UNI 10838:1999. Ente Nazionale Italiano di Unificazione, Milano

UNI (2008) Sostenibilità in edilizia - Esigenze e requisiti di ecocompatibilità dei progetti di edifici residenziali e assimilabili, uffici e assimilabili, di nuova edificazione e ristrutturazione, UNI 11277:2008. Ente Nazionale Italiano di Unificazione, Milano

Vitruvius Pollio M (15 B.C. about) De Architettura [italian edition (2002) Architettura. Biblioteca Universale Rizzoli, Milano]

Weise T (2009) Global optimization algorithms—theory and application, 2nd edn. it-weise.de (self-published), Germany. Available at: http://www.it-weise.de/projects/book.pdf. Last view Jan 2019

Wiener N (1950) The human use of human being. Houghton Mifflin, Boston

Woodbury R (2010) Elements of parametric design. Routledge, London

Chapter 4
Real-Time Monitoring Data at the Time of the Networks

4.1 Introduction

The chapter analyses some possible application scenarios on an urban scale with repercussions on the building scale deriving from the use of large quantities of real-time data produced by low quality sensors spread throughout the territory.

The subject of data on an urban scale aimed at optimising and controlling design is closely related to studies on collective intelligence, techniques and implications due to the management, use and construction of big data, often derived from heterogeneous sources and not only numerical (Mayer-Schönberger and Cukier 2013; Chiesa and La Riccia 2013; Nielsen 2012; Weinberger 2012; Warden 2011; Shirky 2010; Girardin et al. 2007).

Specifically, the implications of the construction and design of real-time monitoring campaigns based on the production of large amounts of data on a small geographical scale are conducted by specialised research centres such as the Senseable City Lab of MIT and the City Form Lab of the Singapore University of Technology and Design in collaboration with MIT. This issue is also addressed by several companies, including Libelium Comunicaciones Distribuidas, which deal with projects on different scales, where the use of low-cost sensors and the production of real-time maps converge in response to the new needs of emerging SmartCities. The topic is directly interfaced with specialised research, conducted on various investigative fronts, to which we refer for specific insights, fully aware of the rapid and consistent evolution of this area, both with regard to DIY systems and with respect to commercial applications.

These solutions are conceived to manage large amounts of information with a view to the development of the city of data and parametric urban planning, in which data, information and technologies take on a real role as parameters of development and civilisation (Vico and Lucà 2005).

The data production nodes of the data city are derived from a vision based on the quantity and almost capillary distribution of the survey points, in which each control unit becomes a node of a network. The set of data produced by the individual nodes

G. Chiesa, *Technological Paradigms and Digital Eras*,
PoliTO Springer Series, https://doi.org/10.1007/978-3-030-26199-3_4

allows scientific analysis to be carried out in specific fields, where the overall error of analysis of small amounts of precise data may be greater than studies derived from large amounts of less refined data (Nielsen 2012). While the single monitoring conducted with sensors and methodologies of the highest quality produces precise data, these are, by their nature, limited in time and space. On the contrary, the dissemination of less refined data production platforms, which are capillary in time and space, allows the construction of a basic analysis that is fed by the large amount of data, modifying the study methods and the implications, also of a design-related and architectural nature. The topic addressed includes the concept of cognitive surplus and intelligence of networks and is directly interconnected with the long tail of information.

To this end, a monitoring campaign was simulated to show how it is possible to develop hardware and software solutions that can be implemented and distributed in the event of a shortage of resources. These solutions can become significant for scientific investigations when the data monitored is traceable and describable. For this reason, the development of the experimental monitoring simulation is accompanied by the introduction of a taxonomy of nodes for the production of urban data that can be implemented and opened, so as to allow work on the construction of metadata, ontology, comparability and formatting of monitored values—see also Chiesa (2014).

The sensors and hardware platforms for rapid prototyping that have been chosen can produce data at different levels of quality, in terms of margin of error, measurement accuracy, resolution, reliability, ease of calibration, ontology of different sources, homogeneity of measurements from different nodes, possibility of having metadata. The sensors used generally perform less well than those used in the laboratory or in on-site measurement campaigns conducted by specialised companies. This is not a limitation of the simulation because the context of application and use is based on different assumptions, although the improvement of boards and sensors directly compatible with rapid prototyping systems, such as Raspberry PI (e.g. Richardshon and Wallace 2013) and Arduino (e.g. Payne 2011), is such as to begin to provide suitable results for scientific monitoring.

The chapter is organised into four steps introduced below.

Section 4.2 deals with the main application issues arising from an approach to monitoring based on the use of distributed data production stations and big data. At the same time, it distinguishes five phases of data: production, collection, analysis, visualisation and use.

Section 5.3 introduces a protocol based on the compilation of definition masks by users for the development and identification of nodes and parameters monitored by the different components of a bottom-up, widespread, low-cost and open source network. The filing of the nodes allows us to guarantee the implementation of the network making the results usable for scientific purposes. The aim is to increase the simple measures that can be implemented online, through a series of metadata capable of rendering the same information.

Section 5.4 describes the results of a monitoring simulation based on the construction of three stations, or nodes, of data collection located in different buildings in Turin. The prototypes were used to record specific variables related to indoor and outdoor air quality following the approach described in the previous point. The data collected have been processed using CAD/GIS software (Rhinoceros, Grasshopper—e.g. Khbazi 2009—and GHowl) for the visualisation of the parameters over time.

Lastly, Sect. 4.5 reports the taxonomy of some of the most common components used to build open source monitoring nodes. The descriptions are divided into thematic areas, such as variables and sensors on the market related to air quality, compatible microcontroller boards and protocols for communication and data storage. The different solutions can be used to populate the station definition mask introduced in Sect. 4.3.

4.1.1 General Objective

The general objective of the analysis of the monitoring simulation presented in Sect. 4.4 is the development of a prototype of a real-time monitoring system to demonstrate the applicability of the emerging DIY technologies to the urban context. The development of nodes in a network of sensors suitable for producing and transmitting large amounts of data in real time allows, on the one hand, the demonstration of the effective validity of these technologies, even by non-expert users in a regime of economic and technical scarcity, and on the other hand, to investigate the applicability of open source technologies and open hardware for monitoring on a bigger scale. The aim is also to verify the applicability of these systems for the smart and real-time design of the city, investigating the problems and repercussions that real-time data and the production of large amounts of data disseminated at territorial level may have with respect to the design and management of the cities of the future. This area is affected by the rapid development and progressive dissemination of commercial solutions and specific research results (e.g. Benammar et al. 2018). The example shown aims to make the phenomenon understandable and, at the same time, provide an easily replicable case study to directly test the potential induced by this innovation.

4.2 Real-Time Data and the Big-Data Revolution

For centuries, the theme of monitoring has been associated with the idea and need to know. Indeed, "to measure is to know" (LORD KELVIN). This concept does not preclude upcoming innovations in data analysis techniques and the new implications due to the spread of big data, especially in cases where real-time data is analysed on small geographical scales. Big data are interconnected with "things that can only be done on a large scale" (Mayer-Schönberger and Cukier 2013) in a context that can be summed up in the word "data", which implies the ability to acquire information and data regardless of the specific initial goals, passing from the collection of a limited number of samples to the accumulation of as much data as possible. Big data differ from small data in that the number of samples tends as much as possible towards the population as a whole.

The implications of more measurements that are more precise in terms of sampling but more uncertain in terms of measurement, affect the aspects of the data, their use and traditional methods of employment. From the point of view of "the thing" and not "the why", scenarios and visions of future use of new techniques for monitoring and data analysis are changing.

There are, however, many applicative implications related to scientific monitoring, particularly to allow the use of the data disseminated, in real time where required, to develop or popularise specific models. These implications are related to the very nature of the data and databases, especially in cases where it is necessary to respond not only "to the thing", but also to some "whys".

4.2.1 Scenarios and Visions

It is now possible to map certain variables in real time, opening up numerous innovative uses of data. Moreover, by using additional monitoring methods, based, for example, on keywords entered in Google searches, it is possible to considerably expand the scope of application. An interesting case in this respect is the study of the spread of the pandemic due to the H1N1 virus in 2009. The dissemination maps elaborated by the *Centers for Disease Control and Prevention*, based on doctors' reports, were able to follow the evolution of the situation only with a delay of one or two weeks. Shortly before, however, Google's engineers had perfected a mathematical method for estimating the spread of winter flu viruses on the basis of a defined set of keywords typed by users that had proved to be strongly correlated with the presence of the virus in the years 2003–2008. The method made it possible to obtain a real time map of the spread of the pandemic which is much more relevant and updated than that offered by the health authorities of the United States (Mayer-Schönberger and Cukier 2013; Ginsberg et al. 2009). The possibility to map the spread of the virus in real time, through indirect correlations, describes one of the innovations and applications typical of the era of big data. Real-time mapping allows not only the implementation of health policies for emergency management, but also the expansion of the possibilities of study and knowledge about viruses and their evolution (Neumann et al. 2009). Of course, there are more and more examples of this.

The ability to map certain variables in real time also has numerous implications in the field of architecture and urban planning. Among the possible application scenarios for big data and real-time data on different scales, the *chip city* concept developed by Hashimoto and Dijkstra, based on the widespread use of GPS systems, is worth noting (Chiesa 2010; Hashimoto and Dijkstra 2004). GPS systems allow positioning, tracking, navigation, map-making and measurement in three dimensions. The chip city is configured as devoid of roads and signalling or advertising, moving these features to a virtual reality control based on GPS and intelligent guidance systems and information management. This research is accompanied by EU-funded research projects into *Intelligent Transport Systems* (see the ITS website) and the various developments relating to driverless vehicles. Currently Google allows individual users to view real-time traffic on major roads and provides information on travel times for specific routes. Previously, some local agencies had already developed systems for real-time traffic mapping, see for example the portal managed by 5T in Turin. On the basis of these projects it is plausible to foresee scenarios in which it is possible to map the positioning of vehicles and traffic, and the urban pollutants present in the territory in real time, and consequently send input to the navigators. This will make it possible to direct users towards optimised routes, in terms of travel time and presence of pollutants for example, with a view to planned management of emissions, thus avoiding congestion or over-concentration in certain areas (Chiesa 2013a). Continuous evolutions in this direction are now evident, even just by checking the potential of the apps of our smartphones. The study of flows mapped by positioning sensors, moreover, allows the prediction of economic development scenarios and market analysis, and new generation positioning. In addition to what has been said, the analysis of real phenomena has multiple direct effects on design on different scales, in the direction of parametric urban planning aimed at the optimised distribution of functions and services.

Also at home automation level (e.g. Wang 2010; for an introduction using Arduino see Riley 2012), it is possible to provide more intelligent management and control systems, capable of optimising the operation and performance of individual units and consumption, introducing any modulations to reduce the risk of peaks and possible blackouts. The implications are considerable with regard to the heating period, as emerged in a conversation with Professor Clarke in 2011 in Turin. The same concept also becomes particularly significant when applied to cooling systems. In fact, most mechanical summer air conditioning systems use considerable amounts of electricity and are responsible for the increase in electricity consumption, peaks and blackouts. The possibility to send specific home automation control systems, which can be managed remotely from mobile phones and computers, inputs aimed at reducing the energy consumed during peak periods by exploiting the thermal inertia of the enclosures and strategies for cyclical switching on of territorial areas based on internal temperatures in real time, could prevent, or at least limit, the risk of national blackout and excessive costs. In this sense, it is also possible to envisage methods for the preparation of such solutions from the design phase of new buildings or in retrofitting actions. Using a massive approach to dynamic energy simulation, it is, in fact, possible to define fits that are capable of linking the consumption or behaviour of the building (e.g. thermal inertia of the envelope) to the design variables—see in this sense (Chiesa et al. 2018, 2019) and Sect. 1.3.3. From the point of view of the operational phase, there is growing research into the topic of smart grids or the results of projects such as SEEMPub (e.g.: Osello et al. 2016).

The application of big data and widespread monitoring networks can also be an important starting point for innovative ways of drawing up emission inventories and energy consumption maps broken down by source at urban or neighbourhood level. The development of the *Convenant of Mayors* initiative promoted by the European

Union has prompted more than 7700 European municipalities to sign up to the Pact and has collected more than 6000 SEAP (Sustainable Energy Action Plans) and more than 2000 monitoring reports (November 2018). Each Plan includes a reduction target of at least 20% CO_2 equivalent and an inventory of emissions in the sectors concerned. The inventory is followed by a list of specific actions that can be monitored over time to reach the desired emission target. The Pact requires that, every four years, the municipalities involved send at least one quantitative report containing a new inventory of emissions and a quantification of the reductions achieved through the actions in place. At the same time, although this is not mandatory for the drafting of SEAP, the adoption of emission reduction strategies can be supported at decision-making level by the development of scenarios of *Business as Usual* and reduction, as suggested by other sources such as the *Cartesio* network and IPCC. The EU, as 2020 approaches, is linking this initiative to the theme of resilience to climate change and urban disturbances, such as the effect of the Urban Heat Island—UHI—e.g.: Yang and Santamouris (2018), Chiesa and Palme (2018), Santamouris et al. (2017), see also Santamouris (2001), Givoni (1998), Nowak (1994), Myrup and Morgan (1972).

Big data and widespread sensors could radically change the way data is collected to draw up inventories and scenarios and, at the same time, provide real-time profiles of disruptive phenomena, from anomalies with respect to typical meteorological data to phenomena such as the aforementioned UHI. It is possible to integrate direct or indirect consumption data subdivided by energy source and those derived from statistical consumption per unit of reference with precise surveys on both consumption by source and actual emissions into the air. Moreover, the adoption of real-time solutions on a large scale based on a myriad of sensors could allow the monitoring not only of the situation on an urban scale, but also the detailed examination of the individual territorial areas, intervening on the most critical issues. The development of solutions in this regard could lead to the development of numerous lines of application including, for example, the need to establish sets of indicators and solutions for overcoming the 2020 target and the *Horizon* programme, achieving the objectives of net positive systems. Current tools should be modified in order to integrate single building visions, context and life cycle assessments, according to a diachronic view, both temporally and spatially. The net positive district, for example, will not be built from the juxtaposition of net positive buildings, but from a broader assessment in which all the buildings will contribute to a local net positive balance. A new building in a given context will have to comply with the needs of the district, to achieve the goal. There are many possible areas of research and development on this subject. A kind of permaculture of energy and different flows.

Moreover, precise mapping of consumption per single object could contribute to the improvement of the accuracy of the LCA analysis in the phase of use of machinery and tools by building new areas of connection between the virtual (model) and the real, looking at new scenarios of overcoming some of the limitations inherent in the formulation of the models themselves (more feasibility, more cost-effective models, more self-adaptability).

Table 4.1, developed in 2012, shows possible spin-offs from the use of big data and real-time information in the urban context, focused on possible applications related

Table 4.1 A sample list of potential applications of big data and IT (first elaborated in 2011)

Topic	Strategy	Potential technologies	Quotation/solution
Intelligent vehicle traffic management	Real time mapping solutions accessible from the internet	Transition sensors located throughout the territory and construction of georeferred real time maps	Data processing at city/metropolitan level (e.g. 5T, Turin) and GoogleMaps/Earth
Intelligent vehicle traffic management 2	Optimization of origin-destination routes using tom toms and navigators based on real-time traffic, transferal times, air quality in different areas including the calculation of the environmental impact of the journey (optimising the related route)	GPS sensors on individual vehicles, hardware platforms to send and receive data, management platforms for individual HUBs and information processing	European research ITS, Chip City (Hashimoto and Dijkstra 2004), Senseable city lab researches
Remote management of single system terminals (e.g. radiators)	Actuators on thermostatic radiator valves and on single-family or apartment building systems for reading and remote management of end-user equipment	Open-source and low-cost software and hardware (e.g. Arduino compatible solutions) or proprietary systems connected to computer systems for remote management of the sensor-actuator system, for example via home internet or using IoT solutions—machine learning systems using phone apps to activate heating/cooling systems reaching setpoint e.g. according to users geolocalisation, behaviour	Solutions developed by makers and available online, proprietary solutions (various companies) in the sector. Possibility to voice-control different sensor/actuators

(continued)

Table 4.1 (continued)

Topic	Strategy	Potential technologies	Quotation/solution
Management of heating and cooling systems for peak attenuation and blackout and consumption management	Centralized management of stop and go and modulation of heating and cooling systems by exploiting the thermal inertia of building envelopes and using internal temperature sensors. It may be useful to have digital models of the building stock on a shared basis to predict the response of the building envelope, such as through BIM or specific software	Sensors and actuators based on specific software and hardware (open source wherever possible) and centralized management, for example via the Dashboard, of network supplies. It is important to establish the nature of the centralized input with respect to the comfort parameters managed by the user	concerning heating district systems (conversation with Clarke 2011)
Integration of models and systems for a connected and accessible management of the building (comfort, temperature, humidity, air quality)	Actuators on supply terminals, air flow management via passive and/or hybrid solutions, integrated flow management (gas, electricity, heat, cold) with centralized production and customized solutions for individual users	Use of low-cost open source software and hardware connected to a data processor for building management, integrated building systems in the technological elements of the buildings to bring the flows to individual users, while leaving ample possibilities to customize management on the terminals (e.g. integrated systems in the coat)	

(continued)

Table 4.1 (continued)

Topic	Strategy	Potential technologies	Quotation/solution
Implementation, potentially in real time, of GHG emission inventories and energy consumption at different scales, in order to map and reduce emissions and consumptions by intelligently managing the various energy sources—Covenant of Mayors, projects for intelligent mobility, BEI (baseline emission inventories) and BaU towards a real-time monitoring of SEAP (Sustainable Energy Action Plan—city level) actions	Development of monitoring and data collection networks to map consumption or improve the methods of estimating consumption by source, in the case of typological classifications. Monitoring of emissions and air pollutants. Monitoring of CO_2 and other pollutants to verify the functioning of the actions envisaged by urban climate plans. This can also be applied to increase energy performance of buildings and/or to suggest real time labelling	Sensor systems, data management and collection platforms. Elaboration of statistical solutions to improve the estimate of consumption for typological classes	
Urban heat island, monitoring and mitigation strategies	Testing of summer overheating of outdoor environments. Identification of design solutions to reduce heat islands	Diffused temperature and solar radiation sensors, including globe thermometers. Implementation of real time and seasonal averages maps, development of intervention protocols. Including pedestrian and user sensations by app-connected devices to suggest priority-intervention maps	Heat Island Group research group of Berkeley Labs and Akbari publications on heat islands and mitigation strategies. Refer also to Santamouris (e.g. 2011 or Santamouris et al. 2017)

(continued)

Table 4.1 (continued)

Topic	Strategy	Potential technologies	Quotation/solution
Development of design tools for post Horizon 2020	Definition of net positive district solutions based on sets of need profiles distributable on the network	Use of parametric evaluation software, implementation of model construction methods for the project and interoperable management between the different softwares, verification on district and building scales of single values. Development of sensor and actuator networks for the intelligent management of local networks in operational rating	
Implementation of dynamic simulation and real-time operation of building systems to maximise the usage of passive and low-energy climatic potential	Definition of a direct connection between real weather conditions, building sensoring and system activation to balance hybrid systems and allow for maximal usage of local natural sources for passive heating and cooling of spaces	Low-cost sensor/actuator systems connected to dynamic energy simulation engines to check for optimal configurations of passive and low-energy systems (e.g. ventilative cooling—natural forces or fan-activation) to maximise low-energy comfort. Machine learning can be considered to directly connect external conditions with building behaviour	See earlier definition in Chiesa et al. (2018)

to the urban monitoring of air quality and flows. Some of the solutions presented here are already partially implemented today, emphasising the speed of innovation in this field of R&D.

4.2.2 Aspects of Data

Innovation in the processes of management and creation of information and data has profound repercussions on the specific technical-application modalities of multiple aspects of data analysis. The choice of dividing the aspects inherent in the use and production of data into different areas allows clarification of different lines of research and the analysis of different implications related to the world of big data and real time. Actions concerning data and information are broken down according to the main reference macro-areas below:

- Direct and indirect data production.

This aspect involves direct data production techniques, such as sensor monitoring, use of simulation software and indirect production techniques, such as data mining techniques and extrapolation of information from large amounts of data created for different purposes, such as market data and preferences from Facebook or Twitter. These techniques also include the processing of datasets for the creation of new derivative fields;

- Data collection and storage

Collection and reorganisation of data in datasets and thematic databases organised for use in specific software. It also includes data storage and distribution techniques, organisation of transmission networks and *cloud computing* and on-line storage techniques. This point also includes formatting techniques aimed at making uneven data processable;

- Data analysis, processing and optimisation.

Data analysis through statistical techniques, modelling of territorial phenomena or those related to individual buildings/plants, accessibility studies, optimisation of processes, forms and distributions, evaluation of performance on the basis of data against specific indicators derived from user requirements, without forgetting machine learning techniques and the possibility of using neural networks;

- Data visualisation.

Creation of real-time maps and computer models capable of interacting with data from monitoring or other sources. In the cases of geographical representations composed of features and their spatial relations we talk about geovisualisation, while we can speak of datascape in cases where 3D visualization uses space as a metaphorical place to make data related to a specific spatial dimension interpretable and/or communicable (Vico and Lucà 2005);

– Data use.

Use of the data analysed for commercial, planning and optimisation purposes, feedback cycles through actuator-sensors, materialisation of information for management of real objects and changes in performance based on the values achieved.

It is necessary, however, to remember how the flow of data from production to final use crosses different areas of influence, characterised by diversified expertise and specific backgrounds, such as network or production nodes, storage and processing platforms and operational scenarios that identify, within the datasets, the components useful for analysis. In the world of big data, it is important, right from the collection of the data points, to make the data structurally flexible in order to facilitate different possible methods of use and aggregation. "The additional cost of collecting multiple data flows, or many more data points within each flow, is almost always low. So it is advisable to collect as much data as possible, and to make it extensible, taking into consideration future secondary uses from the outset" (Mayer-Schönberger and Cukier 2013). The possibility of using data according to different operational scenarios, not necessarily established at the time of collection, with a view to extensibility, leads to an increase in the potential value of such data.

4.2.3 Application Challenges

The development of widespread monitoring nodes for the production of large amounts of data opens up numerous problems of data management and use. An element to be taken into consideration in analysing large amounts of data is represented by Fig. 4.1a, in which we can see how the amount of information is inversely proportional to the legibility of the visualisations. Nevertheless, the assumption of specific scenarios of platform and hub operation enables the solution of this problem

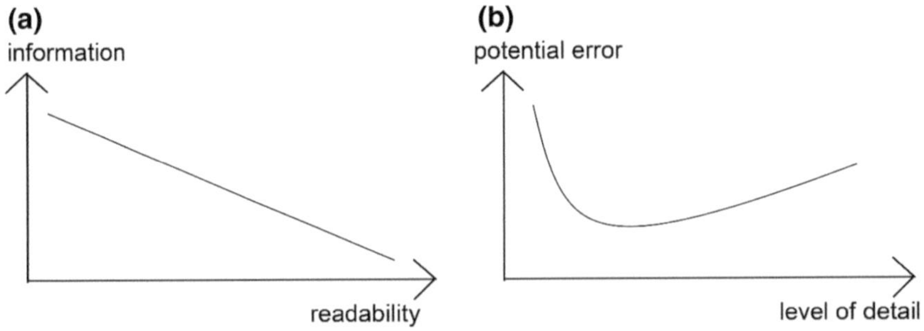

Fig. 4.1 **a** As the displayed information increases, the readability of the representation decreases (Vico and Lucà 2005); **b** when reality is translated into models to be simulated the choices of correct input data and results' interpretation are essential. Increasing details increase also the potential feasibility of results, but after a certain level the total error level starts to increase again due to overfitting phenomena and impact of potential modeller errors (Heiselberg 2018, 66)

by choosing the correlations and information necessary for individual analysis, ex post rather than ex ante in the case of big data. In the case of big data, in a scenario in which the computing capacity is constantly growing, the need to solve the problem through classifications is diminishing. The actual data begins to be able to be processed directly, as can be deduced from the work of the Senseable City lab and the City Form Lab. However, if these data have a possible use or simulative end, it is necessary to remember how the increase in the complexity of the models (e.g. level of detail), while leading to a hypothetically more precise result, leaves room for greater margins of error—see for example that discussed in Heiselberg (2018) in relation to ventilative cooling—IEA EBC Annex 62—Fig. 4.1b.

The main difficulty of analysis related to the spread of acquisition nodes stems from the lack of metadata related to the measurements, i.e. information that introduces, for example, the type of variable handled, the source and the methods of acquisition. Metadata is a fundamental element for moving from data to information because it is essential to know the type of data variable considered, whether it is nominal, ordinal, interval or ratio-based, the definition of what and how it is measured and the operational definition of variables. Finding this information is often a difficulty which can be overcome with some precautions, as suggested in the following paragraphs. Considering the georeferenced or positional nature of the measurements, it is also important to assess problems related to the spatial dependence of the measurements, as much as it is mainly fundamental for small numbers in scenarios affected by the normalisation of data such as, for example, the population variation in a cemetery census plot in which there is only one inhabitant and one dwelling. In the case of big data, the large size of the dataset allows the limitation of problems of a spatial nature by treating the choices of aggregation of data at different territorial levels to verify the emergence of criticality. Finally, in big data analyses, individual datapoints are no longer critical because the system of abundance of information reduces the importance of the individual datapoint compared to the analysis by limiting its ability to pollute the final result (Mayer-Schönberger and Cukier 2013). In this sense, time aggregations can also be adopted to eliminate fluctuations due to excessively specific phenomena (e.g. Acquaviva et al. 2015) for the case of smart grids and district heating.

As regards the quality of data, the following must be taken into account:

- the completeness, in the sense of coverage of the entity involved with respect to the area of study and the maintenance of the information necessary for the classification of the data;
- logical consistency, in the sense of preserving logical relations between data, avoiding contradictions between informative content and ensuring consistency and reliability;
- accuracy, metric precision of the data understood as dispersion of value;
- positional accuracy in the case of geo-referenced data;
- temporal accuracy, understood as temporal reliability (ageing of the data), but also as precision in the identification of the moment of acquisition;
- thematic accuracy, discrepancy between monitored theme and real theme;

– resolution, size of the smallest monitorable value;
– lineage, temporal traceability of subsequent changes.

In the case of big data, however, greater inaccuracy errors are tolerable than in the case of small data scenarios, due to the increase in the size of the dataset. Remember that, in general, it is often more convenient "to tolerate a certain margin of error than try to prevent it (especially in the regime of abundance of information) sometimes 2 + 2 equals 3.9, and this is a fairly good result" (Mayer-Schönberger and Cukier 2013). In other words, we move from the most correct data possible to data that is *good enough*.

Another problem is linked to the ontology of the data, used "to classify and explain the characters of what it classifies" (Ferraris 2003). The methods used to represent knowledge must be selected to suit choices in terms of storage, management, interoperability, transferability and accessibility of data (Rivoltella 2010; Odifreddi 1994). Ontology comprises a conceptual schematisation that translates into the representation of a model of a given domain of an explicit and unambiguous nature, formally expressed in a known language according to a shared knowledge. In a data processing mode based on structured databases, the ontology of data and variables is fundamental in order to proceed with the analysis and comparison of information. The data must be homogeneous and organised so that it can be used and processed. However, tools that operate on large datasets, such as *Hadoop*, a freely distributed computing software, do not necessarily require the extraction, transferral and loading of data for analysis. In other words, "it assumes that the data is not homogeneous and organised - indeed, it assumes that it is too large to clean up before processing" (Mayer-Schönberger and Cukier 2013).

In addition to the issues mentioned above, there are a series of software and hardware problems. First of all, especially when working with data and models, it is necessary to verify their interoperability within the different management and processing platforms. The word is one of the key elements of the use of information technology for design, especially when looking scientifically at the model as a tool for contemporary design. The possibility to access data not only deals with the transferability of the same data from and to different platforms, but also with the persistence over time of compatible access modes. Within the context of widespread monitoring, it is necessary to consider the temporal aspect of the management platform and the individual nodes. Over time, these elements will be updated, with subsequent releases, interface changes and operational tools, and with a view to spatially and temporally widespread monitoring, it will be necessary to ensure the accessibility and operability of nodes and platforms even when they become obsolete. An example is the case of *Pachube*, an open platform for managing the data produced by different types of DIY platform. The acquisition of Pachube by *Xively* has changed the way the platform is used, making the previous connection methods, described in literature and on industry blogs, obsolete and no longer usable. This has resulted in the interruption of the flow of information that specific tools in plug-ins and software made it possible to establish between hardware components and the software platform. Changes of this nature affect the operation and functioning of the widespread monitoring platform, especially in view of the fact that the analysis and collection programmes are organised according to interoperable platforms. When a connection is broken, the flow of information may also be lost. Another example is the case of the Firefly plug-in for Grasshopper. When the new version of Rhinoceros (upgrade from 4 to 5) was released, a dedicated version of Grasshopper became available, interrupting releases for the previous one. The version of Firefly for Rhino 4 made available at the time did not include some of the basic tools that were previously accessible and that were included exclusively in the version for Rhino 5. This makes it necessary to update the CAD software, in the event of system formatting, in order to be able to continue to guarantee the operation of the gate between Arduino and Rhino—at the moment, the current version of Rhinoceros is 6.

The design of a widespread monitoring platform must consider:

- the possibility to guarantee long-term connectivity between the hubs and the platform, both in the case of upgrade/obsolescence of the nodes, and of the network, as well as the platform itself;
- the possibility to guarantee long-term interoperability and functioning of the data flow from production to use/display;
- the possibility to guarantee long-term operation and type of interface, in order to avoid forcing users to make expensive and complicated updates;
- the possibility to guarantee the use of the scenarios following changes in formatting, and for different uses and reuses of the datasets, in order to use the extensibility of the data points collected.

4.3 Urban Monitoring in Smart Cities—The Data-Production Phase

As reported at the beginning of the chapter, Sect. 4.3 onwards looks at the development of a monitoring system aimed at the production of large amounts of data through nodes of sensors located throughout the territory based on open source and open hardware solutions. Once the specific objectives have been identified, a monitoring simulation implemented through three data collection nodes located in Turin is presented (Sect. 4.4)—see also (Chiesa 2013b, 2014). A method for the identification and description of the nodes of the network based on identification cards to be completed by users to increase knowledge of the individual data items produced is also developed, contributing to the construction of metadata and making it possible to analyse and optimise the values monitored. This tool can be implemented and integrated and is useful to make the simulation applicable to real cases. The study focuses on specific data production techniques for air quality monitoring objectives, differing from other big data analyses based on the data-ization of indirect data sources to be analysed, such as Twitter, Facebook and Google. The desire to extend the network of sensors compared to the traditional methods used to sample a limited number of nodes connects reasoning with the big data revolution. The term big refers mainly to the extension of the sample compared to the entire data set in a vision in which N tends towards all (Mayer-Schönberger and Cukier 2013, 47).

The implementation of widespread monitoring systems capable of producing large quantities of distributed data, generally of lower quality than those produced by traditional monitoring stations but inexpensive, implementable and capable of mapping an entire territory, is a rapidly evolving research topic. Numerous institutes, research centres and universities are developing projects related to this topic, including the Senseable City Lab and the Media Lab of MIT, the City Form Lab of the Polytechnic of Singapore, and the IAAC (*Institute for advanced architecture of Catalonia*). The list is constantly growing and also includes municipalities that deal with the development of intelligent systems for the collection and dissemination of information, often implementing systems related to digital islands and augmented reality, as well as businesses and start-ups. The latter include *Libelium Comunicaciones Distribuidas S.L.*, based in Zaragoza, Spain, as a developer of Internet of Things solutions for smart cities that use open source sensors, developing software and hardware packages for monitoring and integrating furniture and urban objects. An interesting example is Waspmote Plug&Sense! (Libelium 2012, 2018a), a design for an urban pollutant monitoring station based on open source platforms and its recent versions calibrated for different types of use—e.g. Libelium (2018b) devoted to smart city solutions; Libelium (2018c) focused on gas detection; Libelium (2016) for event detection scenarios; Libelium (2015) for agriculture scopes.

The platforms designed are linked to data production nodes produced in compliance with the creator of the network itself. An interesting point of innovation is the possibility to design data collection platforms capable of interacting with acquisition stations built by different operators and users, not necessarily conforming to a standard. According to this approach, in order to develop hardware and software solutions for the production of big data related to specific variables, with a view to scientific monitoring based on the long tail of the data, it is necessary to carefully plan the network to allow anyone to increase the database and use it knowing the metadata of each monitoring node. The need to provide a specific description together with each measurement, requires some precautions. We cannot presume to be able to control every data production node, but it is possible to build a common base. Standardisation can be carried out using specifically developed tools, such as tables that describe the individual nodes, supplied by the network operator and written in such a way as to be implemented by individual users with the help of input masks. This solution can be used to build a community of monitoring stations according to an approach that adds a minimum amount of ontology of the data transmitted to each node. Moreover, filling in a form makes it possible to define the metadata associated with the variables monitored. The aim of the process is to achieve the homogeneity of the big data collected in order to subsequently process high-quality analyses and optimisations. The expression "information = data + metadata" is used.

4.3.1 *Identification and Description of the Nodes of the System—Stations' Catalogue*

The use of open source platforms for real-time monitoring on different scales is linked to the type of microcontroller board used, to the sensors connected to it, to the choices in terms of data management and production that can be set by compiling the software, and to the data transfer and storage methods. Within the scope of the survey, the decision was made to describe the boards used and the sensors connected, using an approach that can be reproduced and increased over time. This choice makes it possible to connect diversified nodes to a hypothetical cloud. These nodes can be assembled according to an open approach and are recognisable by the specifications reported by users on a definition mask. As mentioned earlier, it is essential to respond to the application problems related to data ontology, the definition of metadata and the possibility of comparing the variables monitored. This is why it is necessary to be able to set some simple application standards to manage large quantities of real-time data produced by different nodes. It is also believed that the compilation of a station card by individual users can increase the correct identification of the data monitored and the quality of metadata related to the variables monitored in the specific node.

4.3.2 *The Node-Definition Window*

The ultimate goal of monitoring stations connected to a monitoring platform is the widespread production of data. In order to achieve this goal, it is necessary for every user who intends to connect a station to the cloud, providing monitoring data, to fill in a form that describes both the station and the individual sensors connected to it.

Each node is described according to the type of microcontroller board used, the sensors connected to it, the geographical location, consisting of the coordinates and identification of the position—outdoor, indoor and building type—and the identification of any mobile nodes, the start date and shutdown date where known, the additional expansion boards, the methods of transmission, data storage and the type of power supply. Each variable is described by the sensor used and by the techniques used to produce the data, such as the acquisition range and the weighing between several measurements. Thanks to this information, it is possible for the platform administrators to carry out searches on different fields, such as the location, the type of microcontroller and the variable monitored, reorganising the database available. The compilation of the card is used mainly to automatically analyse the various types of data. The user can be directed to the description of the node and the variables monitored thanks to simplified masks. Another element aimed at the construction of metadata is the possibility to insert photographs of the node and its positioning and to attach the relative hardware diagrams and software lists.

Lastly, it may be useful to allow the manual insertion of files with different extensions containing data monitored by stand-alone systems recorded, for example, with the help of SD cards or other methods.

In order to allow the individual nodes to easily transmit the data monitored on the platform, specific software lists and methods of writing the results must be provided to facilitate the implementation of the databases. It is essential to direct users towards a guided standardisation of output management.

To make it easier for users to compile station-cards and variable-cards, it may be useful to provide formats with drop-down menus. However, it is necessary to allow the integration of lists on the basis of specific hardware components not yet classified, such as microprocessor boards compatible with the system, but produced by other companies, or sensors that have not yet been boarded.

The user is required to fill in the following fields in relation to the single monitoring node:

- measurement start date;
- final measurement date, if known;
- method of temporal definition of the time and date of the individual datapoints, such as RTC modules or other solutions;
- georeferencing of the station, which can be carried out using an interactive map, by entering coordinates or thanks to the identification of GPS cards. This data must be entered correctly in compliance with the specifications in the monitoring outputs. Mobile stations such as those located on transport vehicles, must be connected to a geographic localisation module;
- identification of the type of measurement (outdoor or indoor—and type of building);
- type of microcontroller board;
- data transmission mode;
- data storage mode;
- any other cards or functions connected in addition to the sensors;
- number of sensors;
- station hardware diagram. This can be done by attaching files produced using specific software such as *Fritzing* (*.fzz file), drawings created especially or through a system implemented directly in the platform;
- insertion of photographs of the hardware and the location of the platform;
- insertion of the software list.

Each sensor connected to the node can be described by the points listed below, respecting the order of insertion and the specifications supplied for the preparation of the outputs by the manufacturers and operators of the platform—of course some of these points may be redundant and derived from the identification of the sensor:

- variable(s) measured;
- brand and/or type, also by means of an identification code, such as DHT11;
- power supply, variable depending on the type of sensor;
- connection pin;

- acquisition range;
- use of data weighing modes on several acquisitions, such as the average of n values or the mobile average;
- type of transmission and storage, which may differ from other sensors connected to the same node;
- unit of measurement and formula for translating the values into mV (varies depending on the resistance installed);
- positioning (indoor, outdoor and photograph).

In conclusion, the node must be connected to the platform in order to record the data monitored.

As mentioned above, the collection and pre-processing of data monitored is a characterising element of the various sensors connected. The data produced during monitoring can be distinguished according to the methods of collection and pre-processing. In order to describe the value monitored, it is important to first define the interval between the different acquisitions, which can be modified in the software list by using the *delay(ms)* function without forgetting the waiting periods inside the list loop. Data can also be distinguished by weighing or alignment of the measurement. The following can be recorded:

- instantaneous measurements supplied by the sensors;
- averages of a defined number of measurements;
- mobile averages over a defined number of recent measurements;
- mobile averages weighted over the time interval of the measurements.

The information reported in Sect. 4.5 provides a first articulation of the drop-down menus that can be implemented during the development of the nodes and the network. Individual nodes can be built for monitoring use only or can be integrated with other local node functions such as home automation systems and other domestic or public electronic components scattered throughout the territory.

It is also possible to flank the platform with various activities to support the development of metadata and individual nodes, consisting, for example, of specialised blogs and software and hardware kits to be distributed in compliance with the logic of "give away the bits and sell the atoms" (Anderson 2012) and typical software lists for the most common sensors.

The example given in the following paragraph focuses mainly on aspects related to the monitoring node, the HUB, leaving out platform specifications and operational scenarios. In the simulation, the platform is simplified by a direct link between dataset and geometric software, maintaining the georeferencing of the individual system nodes.

In order to develop a complete system for the real-time monitoring and production of large quantities of data spread over the territory, it is important to develop specific platforms for data collection, storage and processing in addition to the individual nodes and system hubs. The platform-hub system can also be used according to different operational scenarios. The scenarios are a point of analysis to be developed in subsequent research work on these topics.

4.4 Simulation of a Potential Monitoring—Description, Analyses and Results

Sect. 4.4 sums up monitoring simulation based on three data collection stations located in different positions in Turin. The three stations, installed to monitor air quality, have similar characteristics in order to be able to compare the data recorded at the same time by the different nodes. The objective of the simulation is to develop widespread monitoring systems that are self-financed and receive no specialist support from electronics experts to demonstrate that it is possible to disseminate the development of microstations to an interested audience. The aim is to create a real-time map simulation capable of reading the data recorded by the three stations (consisting of five variables per station) and bring them back into an urban context by displaying them on the basis of the passage of time. Rhinoceros v5 SR 6 software and the Grasshopper and GHowl plug-ins are used for the display. The data was transmitted via Excel spreadsheet. It is possible to use systems directly connected with google maps, such as Pachube, now Xively, capable of displaying data in real time on georeferenced locations (Faludi 2011). A list of numerous software products and solutions related to big data is given in Warden (2011).

Section 4.4.1 describes the hardware components of each monitoring station and the software developed specifically for the acquisition of data. Section 4.4.2 describes how the data is processed and displayed.

Please see Sect. 4.6 for comments and conclusive details. See also the broader discussion on the topic in (Chiesa 2014, 2017).

4.4.1 Single Node Description

The three monitoring stations based on the same sensors used during the simulation were located in three different places in the city of Turin, in indoor contexts. The different locations are distinguished by method of use: a residential building (station I), a university office building (station II), a university test and component laboratory building which, according to the monitoring results, was found to be a poorly heated room (station III) (Table 4.2).

Following a preliminary test to verify the power supply of station III with AA batteries, an electric socket power supply through 5 V USB adapters was chosen. Two of the stations (II and III) record data on a micro SD card housed in the Arduino wireless SD expansion board. The third station (I), on the other hand, sends the data directly to a computer, recording them in Excel using the macro PLX-DAQ.

The breadboard view of the hardware components of the monitoring stations with MICRO SD recording is shown in Fig. 4.2. Figure 4.3 shows the view for the node connected via USB to a computer. Figures 4.4 and 4.5 show the hardware diagram described according to a conventional display of the station connected via USB (I) and the photograph of the node type, respectively.

Table 4.2 Description of the three test stations—measurements started on 26 November 2013

Station	I_home	II_office	III_laboratory
Geopositioning	45° 04′ 51″N 7° 39′ 17″E	45° 03′ 12″N 7° 41′ 06″E	45° 03′ 44″N 7° 39′ 24″E
Measurement location type	Indoor—residential	Indoor—office	Indoor—laboratory
Microcontroller board	Arduino UNO R3	Arduino UNO R3	Arduino UNO R3
Connected sensors	MQ3; MQ5; MQ7; DHT11	MQ3; MQ5; MQ7; DHT11	MQ3; MQ5; MQ7; DHT11
Power supply	by PC using USB	By electrical grid—USB adapter	Using batteries (failed) by electrical grid (USB adapter)
Storage and data transmission	PC—serial/USB	Micro SD—stand alone	Micro SD—stand alone
Clock reference and data	PC clock—using PLX-DAQ	Arduino clock, by fixing the measurement starting moment—ref. on starting time	Arduino clock, by fixing the measurement starting moment—ref. on starting time
Recording time (start)	08:45 a.m. Day I	11:33 a.m. Day I	09:46 a.m. Day I
Recording time (end)	09:06 a.m. day IV	09:40 a.m. day IV	10:25 a.m. day VII

In addition to the definition of the variables and the inclusion of the SD library, the station software list includes the listings for the operation of the four connected sensors and the controls for writing the results on serial/USB and/or micro SD card. The construction of a variable *text string* was chosen to compose the outputs of the four sensors and send them to the serial and to the micro SD card. This variable makes it possible to read the data recorded in Excel using the macro PLX-DAQ.

The three gas sensors of the MQ series used are characterised by an easy to compile list that envisages the reading of the value on the connection pin of Aout. In the case of the DHT11 sensor, the code is more articulate and includes on/off cycles for reading the relative humidity and temperature values. There are some libraries designed to make it easier to use. The list was put together by entering the complete code. The datasheet of the sensor can be found at http://www.dfrobot.com/image/data/DFR0067/DFR0067_DS_10.pdf (viewed on 04/01/2014), in Chinese, or https://www.microbot.it/documents/mr003-005_datasheet_en.pdf (viewed December 2018) in Italian.

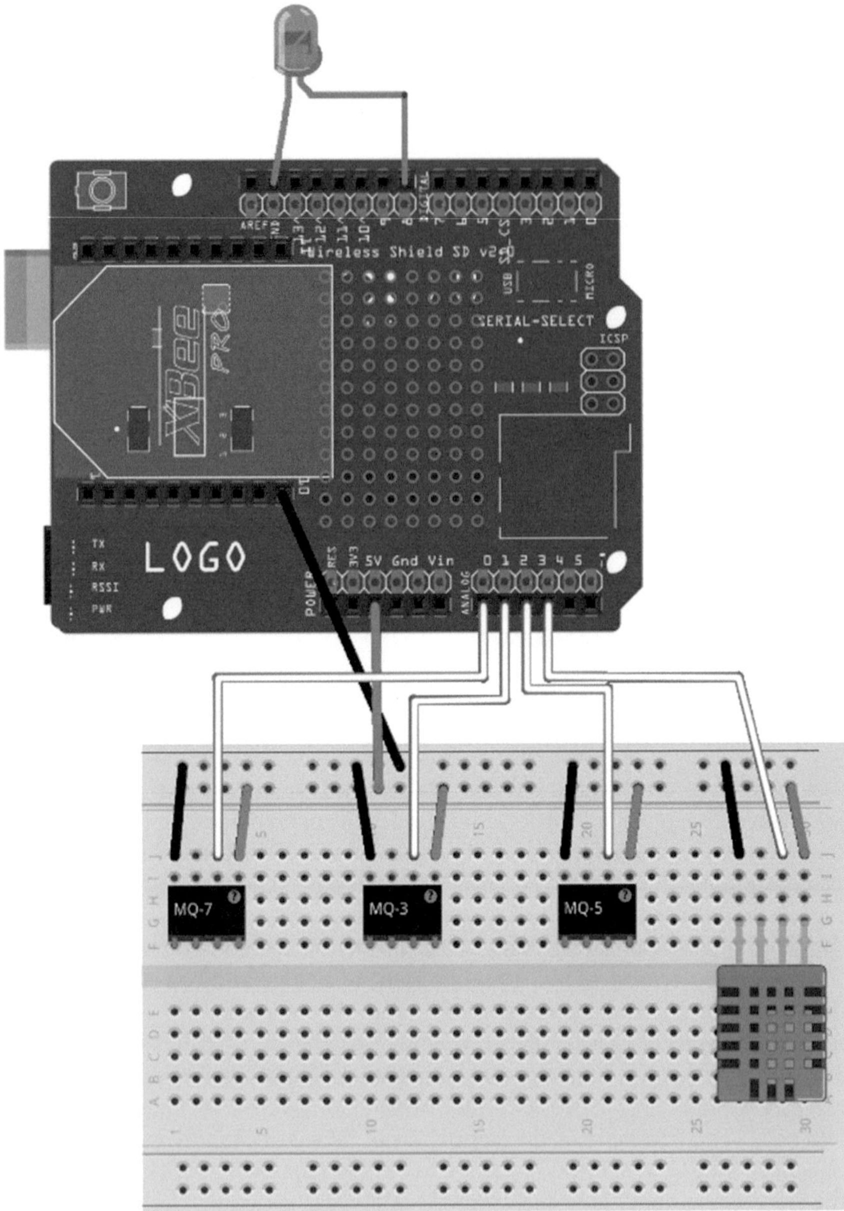

Fig. 4.2 Breadboard view of hardware connections of a monitoring node for the Arduino wireless SD case—schematic view defined by using the free-software Fritzing

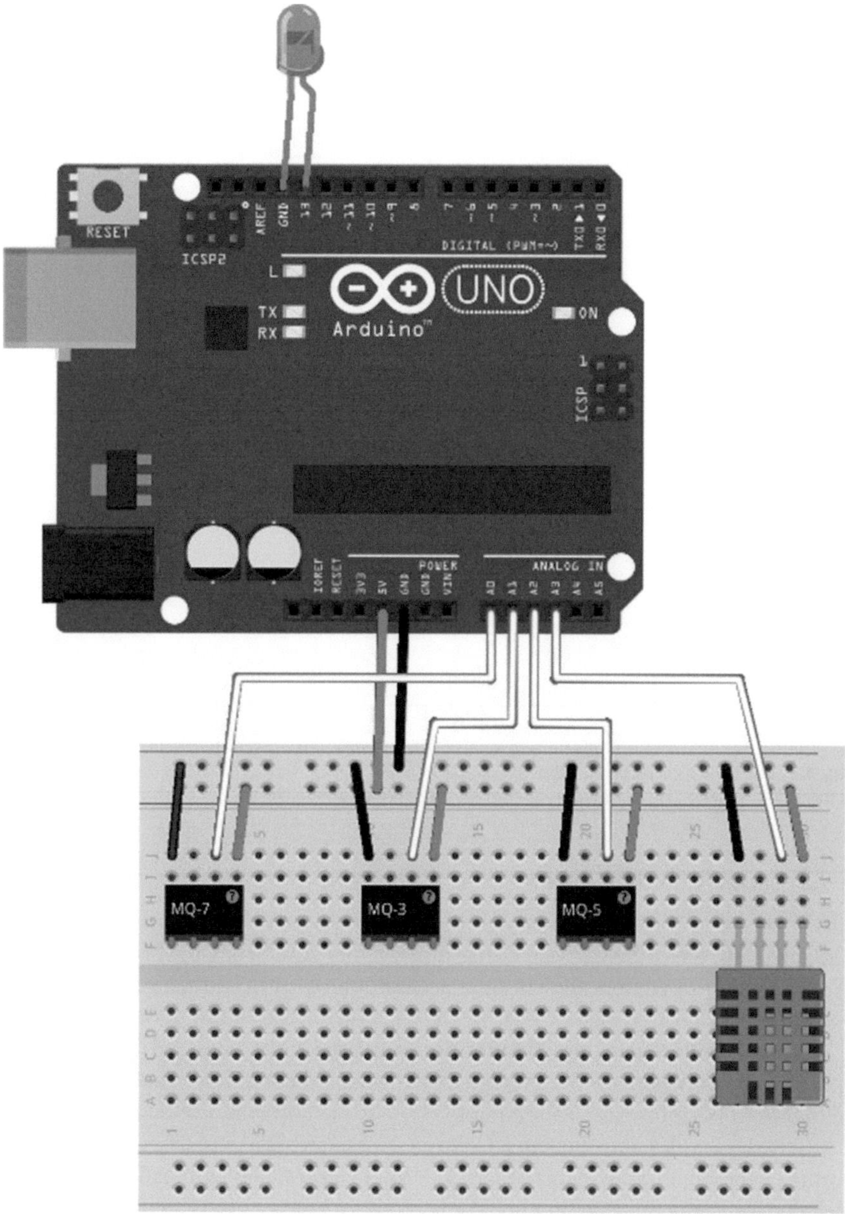

Fig. 4.3 Breadboard view of hardware connections of a monitoring node for the case in where Arduino is directly connected to the PC—schematic view defined by using the free-software Fritzing

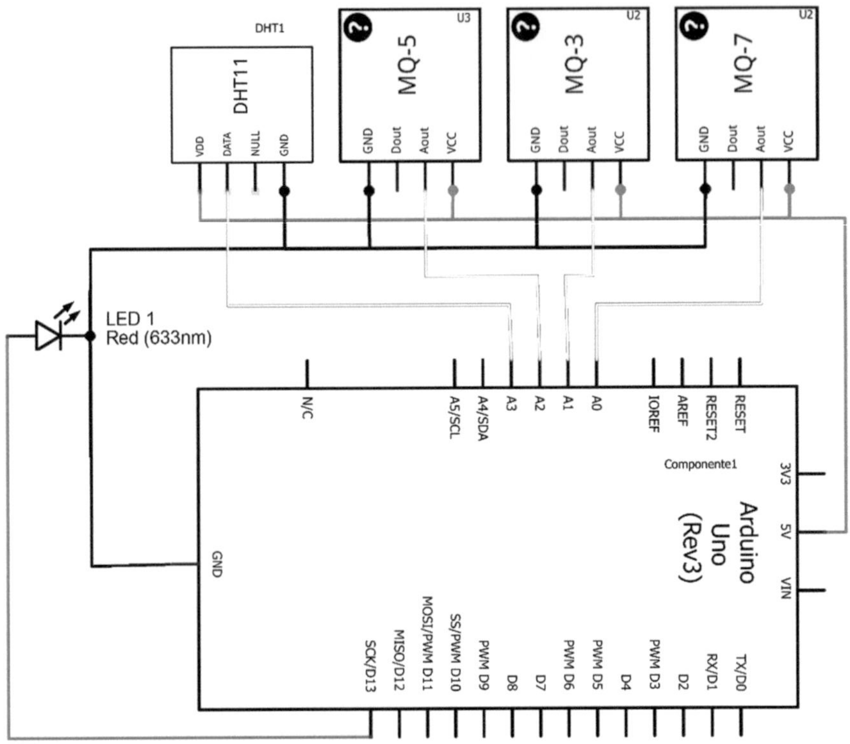

Fig. 4.4 Hardware connection scheme of a monitoring node when Arduino is directly connected to the PC (serial/USB)—defined by using the free-software Frizting

4.4.2 *Analysis of Results*

The three stations were activated after a few hours and turned off after a few days of consecutive operation allowing the acquisition of sufficient data to verify their operation.

The data collected in the monitoring simulation were analysed in Excel. The first step was to calculate the average of the values recorded over one minute. It is advisable to perform this calculation directly at the software list level uploaded in the individual nodes, introducing an *array* for example. It is possible to calculate the average later, creating an Excel Macro in Visual Basic capable of adding up the data included in a given range of time and then dividing it by the number of data items. This procedure may be necessary for stations that are configured differently but connected to the same platform. The number of recordings per minute made by the three nodes should be around 28, as the operation of sensor DHT11 envisages on and off cycles that increase the delay set in the software loop by a few milliseconds. Figures 4.6 and 4.7 show the average measurements for the minute. The values recorded by the three gas sensors (MQ-3, MQ-5 and MQ-7) were not converted into parts per million (ppm)

Fig. 4.5 View of a monitoring node with additional micro SD board

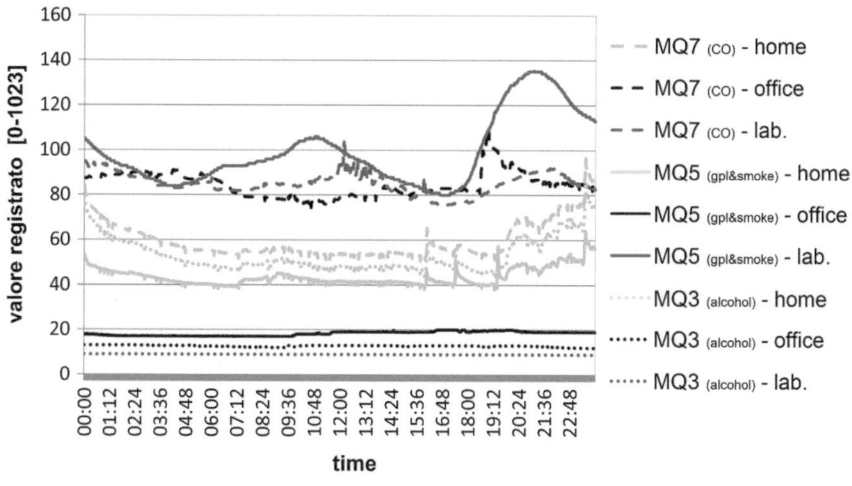

Fig. 4.6 First elaborations of monitored data. Datapoints collected by gas sensors—monitoring day II

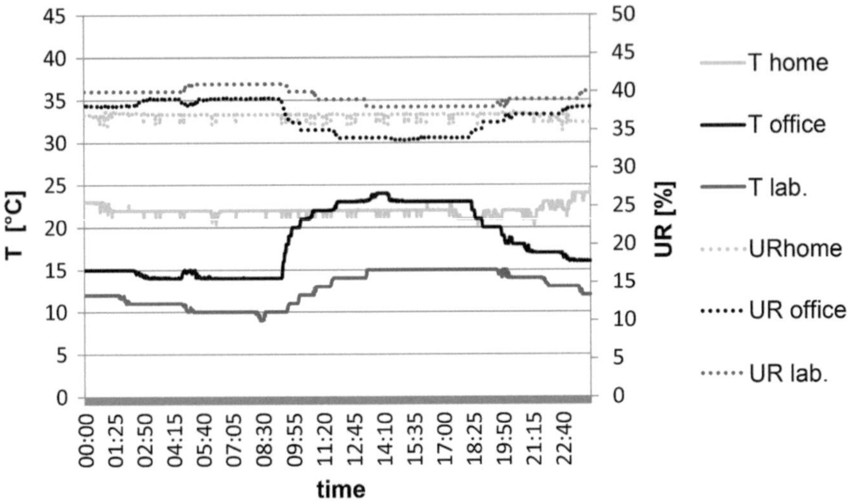

Fig. 4.7 First elaborations of monitored data. Data points collected by the DHT11, temperatures are reported on the principal vertical axes, while RH% on the secondary one—monitoring day II

for simplicity, so they are presented on a scale of 0–1023. Multiplying this value by 5/1023 gives the equivalent in Volts. However, please refer to the information on the manufacturers' datasheets and to the construction of a conversion formula based on measurements in known gas concentrations to make the conversion to ppm. Analysing the graph in Fig. 4.8 it is possible to identify the period of settlement of the measurement, represented by the descending curve traits. The sensors were

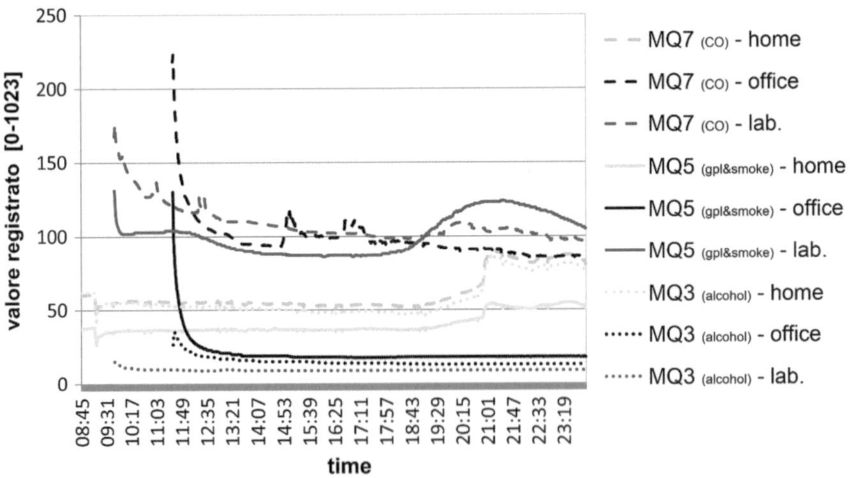

Fig. 4.8 Monitoring day I, gas sensors database. See the adjustment period due to the sensor pre-heating phase

preheated for 48 h, reducing the settling time to the first 2–3 h. It is important to take this into account in order to avoid using ineffective measurements. The sensors respond well to environmental factors and are able to record deviations due to the opening of windows. The CO peaks recorded by the sensor located in the office are probably caused by smoke from the boiler chimneys located in front of the building. Figures 4.9 and 4.10 show how the trend of pollutants in the laboratory building

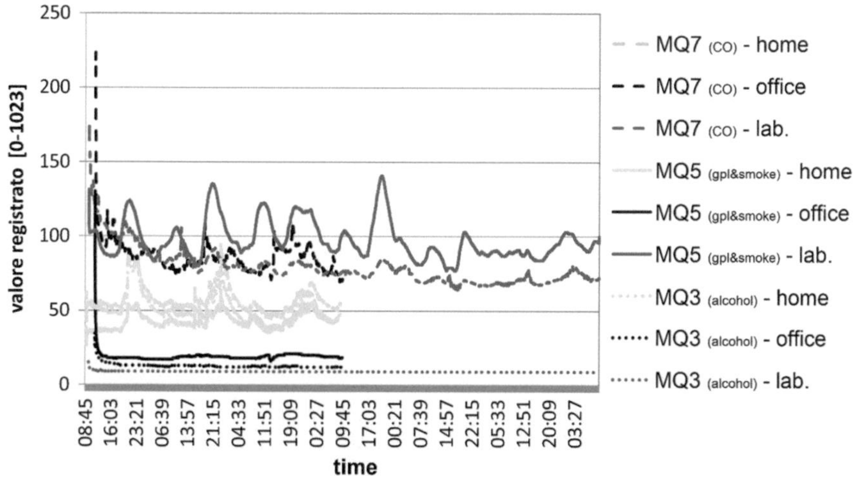

Fig. 4.9 First elaborations of monitored data. Datapoints collected by gas sensors—entire monitored database

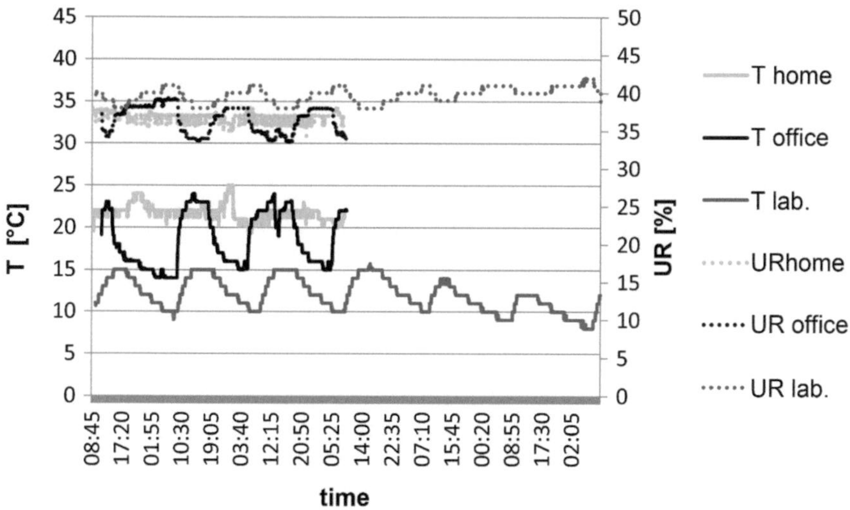

Fig. 4.10 First elaborations of monitored data. Data points collected by the DHT11—entire monitored database. See the cyclical behavior of temperature in office and laboratory buildings

follows an hourly trend, probably due to vehicle traffic at the nearby road junction and the flow of cars in the carpark in front of the building. There is a time delay due to the time it takes for the pollutants to infiltrate the building. It is also possible to compare different temperature and humidity trends. All three stations monitor the trend of indoor environments subject to sensitive heating systems without treatment of the latent component, which is why the relative humidity varies with the change in indoor temperature. The temperature follows a constant trend in the case of the residential building, equipped with central heating without thermal valves, automatic operation in the case of the laboratory, where temperature changes are linked, on one hand, to solar gains, as highlighted particularly by the values recorded on Sunday (last peak of the graph), accentuated by the light structure of the building (sandwich panels), and on the other hand, by a heating system and, lastly, an hourly trend in the case of the office, despite the fact that the switching on and regulation of the heaters is left to the users.

After verifying the operation of the sensors, the values recorded were geovisualised, reporting a timed reading of the datasets on a map. The purpose of the simulation was to verify the possibility to read data from different stations located in georeferenced points in real time. The Rhinoceros V5 SR6 platform and the Grasshopper plug-in were chosen to view the data.

The basis of the display is the geometric component of the shapefile of Turin, supplied free of charge by the city's Territorial Information System (SIT) and downloadable from the geoportal. The monitoring stations were located on the map using coordinates. The script was then processed on Grasshopper to connect the Excel file containing the data monitored to the geometric base. A specific time component of GH was used to ensure a reading of the different values depending on the time and day indicated, obtaining it from the single Excel files connected to the three data collection nodes.

The datasheet was connected to Grasshopper using the GHowl plug-in developed to manage different types of connections to and from third-party programs, including Google Earth. This plug-in is free to download from the internet and can be connected with a timer to time the update of the readings. By replacing the script component that allows the connection with Excel, it is possible to read data from other sources such as, for example, a monitoring card connected by serial/USB through Firefly, or an IP address and MAC using the local or global network, or sites such as Pachube (now Xively).

Figure 4.11 shows the specially developed script and highlights its logical components.

Figures 4.12 and 4.13 show the values monitored at the three stations at two different times (10 a.m. and 8:30 p.m.) on the second measurement day—see also (Chiesa 2013b). In addition, a video was developed using the free software CamStudio V2.7 and VirtualDub V1.10.4 (GNU general public license). As mentioned earlier, this display makes it possible to highlight different phenomena depending on the change in the values monitored, such as the opening of a window, people entering the house, turning off the heating in the office.

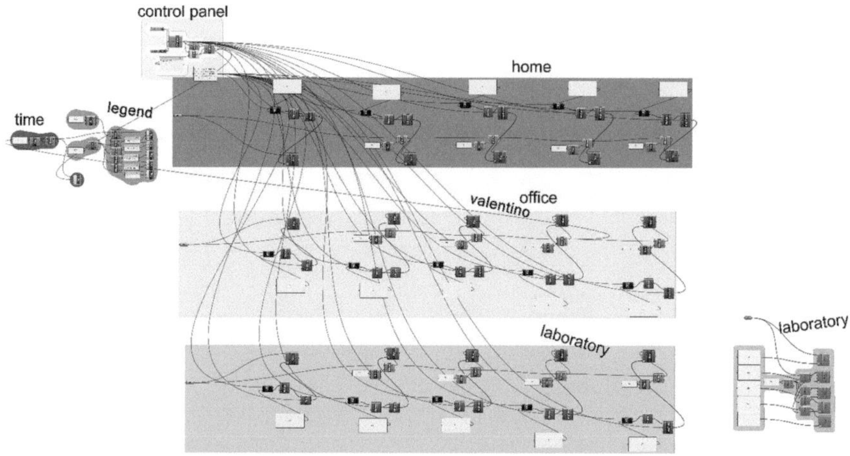

Fig. 4.11 The developed Grasshopper script. Components allow to visualise the monitored data by different stations in a given time

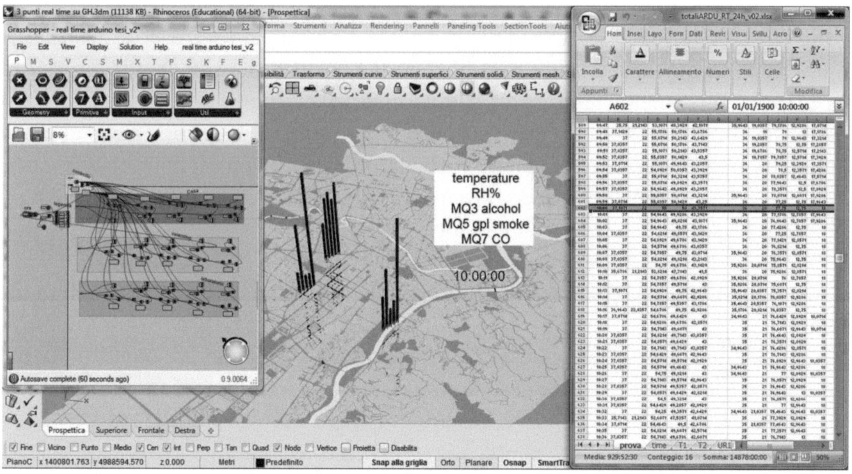

Fig. 4.12 Representation of the 15 monitored values (5 for each station) at 10:00 a.m. of monitoring day II

4.4.3 Further Applications

The operation of a node capable of transmitting the results of the monitoring activity in real time via Ethernet was subsequently tested. The specific aim of this second experience was to validate a different data transmission method. The node was equipped with three sensors: one temperature sensor and two gas sensors, Figaro TGS2602 and Figaro TGS2442. Figaro sensors, unlike the MQ sensors used in the previous example, require more attention during the development of the software,

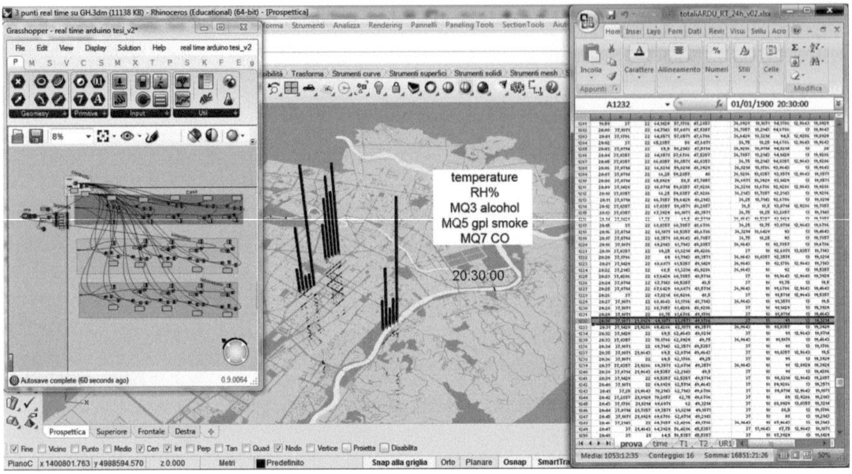

Fig. 4.13 Representation of the 15 monitored values (5 for each station) at 20:30 a.m. of monitoring day II

as highlighted in Sect. 4.5, and in the preparation of the hardware. It is possible to adjust the accuracy of the values recorded on the basis of the anticipated amount of gas in the environment (ppm). This can be done when building the circuit of each sensor, by correctly choosing the resistance to be used. The sensor's reading range is altered on the basis of this choice, working between 0 and 20 ppm or between 0 and 2000 ppm, for example, and consequently the accuracy recorded by the single values of the scale 0…1023 is altered too.

Table 4.3 shows the specifications of the node used.

Figures 4.14 and 4.15 show the breadboard view and the hardware diagram of the node.

The software list includes the SPI, ethernet and SD libraries, in order to use the appropriate tools. The Ethernet card used requires the definition of the MAC and an IP address, at software level. In the case of corporate networks, the IP address must be verified with the managers in order to ensure the operation of the device if there

Table 4.3 Description of the monitoring node connected through LAN

Station	LAN station
Geopositioning	45° 03′ 12″N 7° 41′ 06″E
Measurement location type	Indoor—office
Microcontroller board	Arduino UNO R3
Connected sensors	Figaro TGS2602, Figaro TGS2442, DFRobt LM35
Power supply	by PC using USB or electrical grid using USB adapter
Storage and data transmission	Ethernet board + micro SD

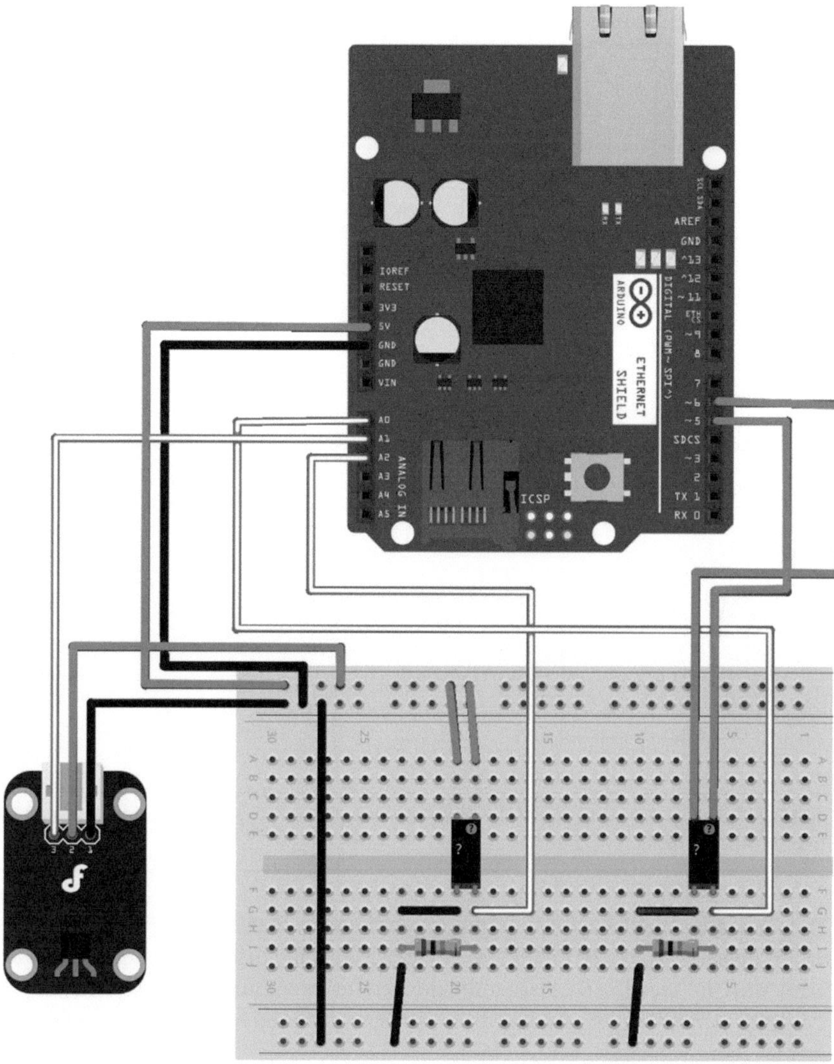

Fig. 4.14 Breadboard view of the LAN station—created by the free software Fritzing

is a block on unrecognised IP addresses. The results or the information transmitted through Ethernet can be read by any device that is connected, whether it is part of a local network, if the IP address works only in LAN for example, or connected worldwide, if it is visible on the Internet, typing the IP address used into the web browser followed by "/?out=all".

Figure 4.16 shows the values monitored by the node on the screen of a computer connected in another room in the building. Figure 4.17 shows the monitoring station.

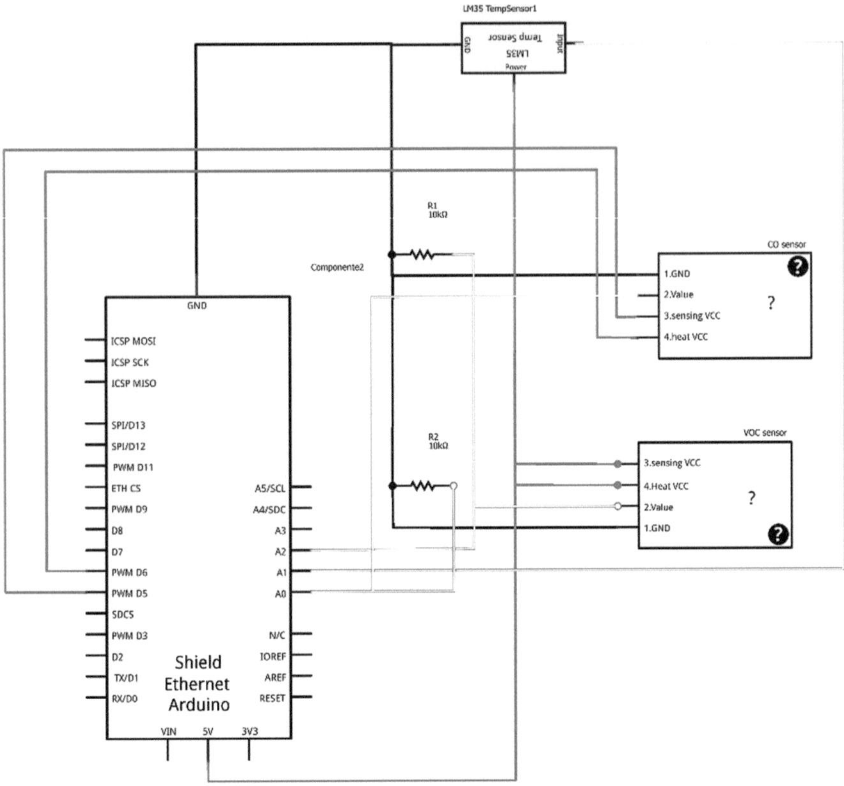

Fig. 4.15 Schematic representation of the hardware connections of the LAN station—created by the free software Fritzing

By connecting the USB/serial output to a computer, thanks to software programming it is possible to see the number of devices connected to the node, i.e. the number of users who are viewing that transmitted by the ethernet node, the operation of the SD card for storing the results and the information related to the connection (MAC, IP, gateway and subnet).

The development of the monitoring station made it possible to verify the operation of data transfer modes in real time via local networks or the Internet. Moreover, the choice of different sensors stresses the importance of the filing introduced in Sect. 4.3.

4.5 Main Involved Parameters—Nodes for Data Production and Collection

This paragraph contains a list and description of some of the most common hardware components for the construction of widespread and bottom-up monitoring stations.

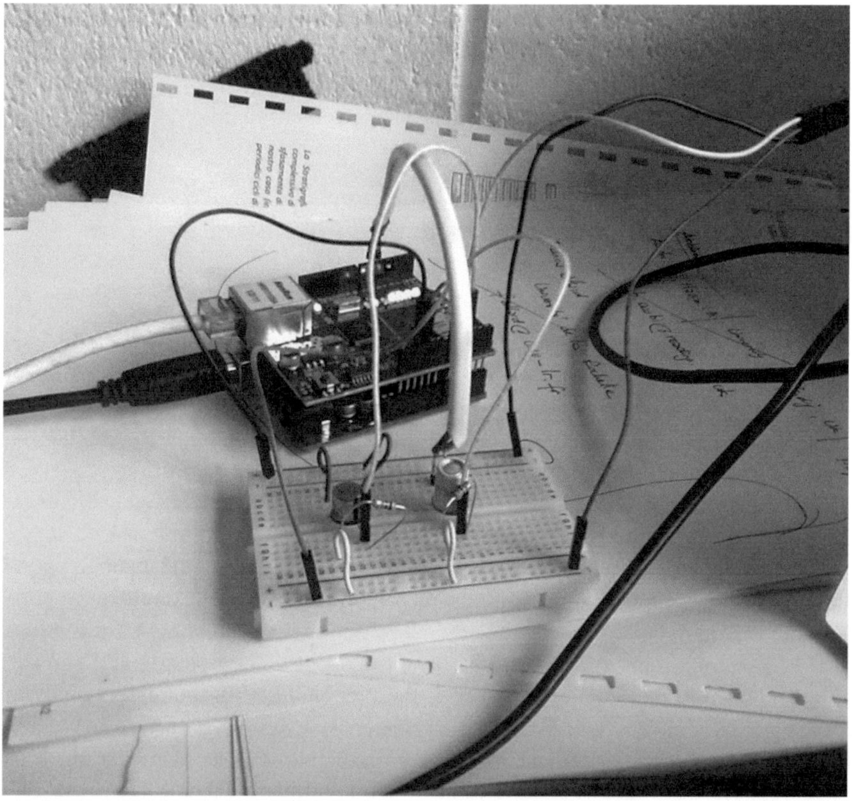

Fig. 4.16 View of the monitoring LAN station

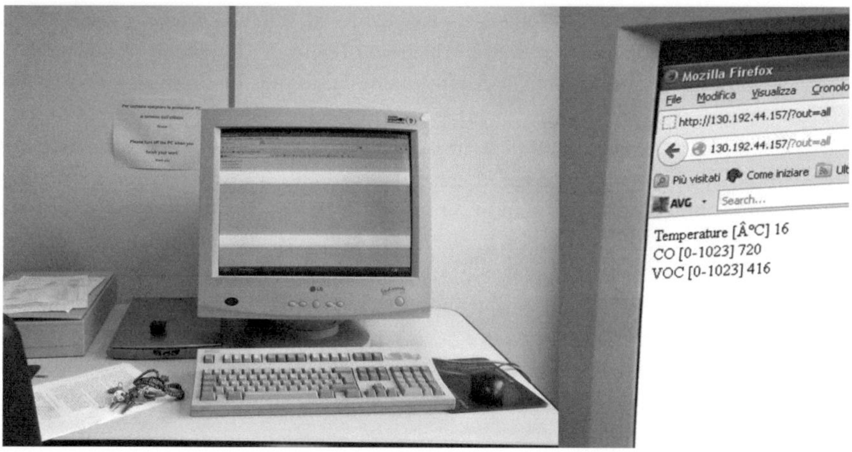

Fig. 4.17 Visualisation of real time measured data on a PC desktop connected via LAN by using the department network. The monitoring node is in located in another office

The list of sensors, microcontroller boards and different methods for the transmission and storage of the data produced aims to define a general framework for the simulation of monitoring described in the previous paragraph and to suggest possible components for the preparation of the station and sensor boards.

4.5.1 A List of Potential Variables and Connected Low-Cost Sensors

The electronic platforms discussed in this chapter allow the monitoring of a large number of variables by modifying the sensors used and the resulting software and hardware schemes.

Before presenting a list of sensors that are easily available on the market, we need to think about the measurement limits due to the resolution of the processing card and the sensors. This concept is important to correctly plan the logical flow and the software and hardware components of the nodes. The Arduino board UNO R3, for example, is a system characterised by a 10-bit resolution on the analogue ports, which corresponds to 1024 intervals, and an 8-bit resolution on the digital PWM ports. Arduino DUE, on the other hand, is a 12-bit system which offers higher resolution.

It is also important to know the hardware connection complexity and software processing of the individual components, their availability on the market and their compatibility, to ensure an adequate power supply, for example. Some sensors require specific external power supply boards because they require a more intense electrical current for pre-heating. The same applies if several probes are connected to the same board. Higher electrical potential values may also be required.

The data monitored by the sensor must be translated into a value corresponding to the unit of measurement of the variable considered, such as °C, % or ppm, using special formulas that can be supplied by the manufacturers' datasheets or deduced by taking measurements under known environmental conditions (Premeaux and Evans 2011).

For example, a measurement on a 10-bit pin reads a value in volts corresponding to one of the 1024 possible voltage ranges in the domain from 0–5 V, approximating the maximum potential of the Arduino pins of 4.995 V. The minimum measurable pitch is therefore 4.887 mV. It is, however, possible to use lower input voltages, connecting 3.3 V outputs, for example, to increase the resolution of the single step, calculated, in the case of 10-bit pins, using the Aref/1023 ratio. If Aref is worth 3.3 V, the minimum measurable pitch would be 3.226 mV. The value in volts measured in this way depends on the resistance value assumed by the sensor when the variable in question changes. This value can be converted into the specific unit of measurement using a special formula that varies depending on the sensor and calibration.

The list of sensors directly compatible with Arduino boards is constantly increasing. For example, it is possible to measure relative humidity, temperature, atmospheric pressure and concentrations of gases such as LPG, methane, illuminating gas, alcohol, smoke, carbon monoxide, ammonia, hydrogen sulphide and weak concentrations of VOC. There are also sensors for measuring geometric distances, inclinations, radiation, bending values, flow rates, contact pressure current intensity, sound waves, heat flows, magnetic fields, biometric values and positioning. Arduino boards, such as the Arduino UNO R3, are not equipped with a real-time clock. There are, however, several ways to flank measurements with the real date and time, including RTC (real time clock) modules that can be powered by battery.

The list of components presented below is a rough guide only and contains some of the sensors tested directly.

Gas and pollutant concentration sensors

The measurement of gas and pollutant concentrations is carried out, at the previous points, using sensors from the MQ series and FIGARO sensors. It is, of course, possible to connect other types of sensor, whether they are directly designed for use on Arduino or Raspberry Pi platforms or have been developed for other applications or systems, subject to the need to ensure adequate voltage and current intensity, as well as correct hardware and software design.

The MQ sensors used to measure gas concentrations, in particular, consist of a heating circuit and an electrochemical sensor. The can be calibrated with a certain precision in the presence of known concentrations of the gases to be monitored, varying the resistance between the output line and the GND. This resistance can vary between 2 and 47 kΩ: low values correspond to a lower sensitivity of the sensor, while high resistance values make the sensor less accurate in the presence of high concentrations of gas but allow its use at low concentrations. The sensitivity of the sensor can be adjusted using a potentiometer after creating a curve on the basis of known concentrations. The output of the MQ sensors is analogue. These components have six pins and are symmetrical with respect to a central axis, meaning that they allow the inversion of pins A and B (input-output). They require a long preheating period of at least 24 h to achieve a stable measurement.

The MQ sensors used in the chapter are:

MQ-3

- Gas: high sensitivity to alcohol, low sensitivity to petrol
- Power supply: 5 V
- Interface: analogue

MQ-5

- Gas: high sensitivity to LPG, methane, illuminating gas; low sensitivity to alcohol and smoke
- Power supply: 5 V
- Interface: analogue

MQ-7

- Gas: high sensitivity to carbon monoxide (measuring range 20–2000 ppm)

– Power supply: 5 V
– Interface: analogue

The FIGARO sensors used are (Figs. 4.18 and 4.19):

Figaro TGS2602

– Gas: high sensitivity to VOC and odorous gases (ammonia, hydrogen sulphide)
– Power supply: heating circuit 5 ± 0.2 V direct or alternating current; measuring circuit 5 ± 0.2 V direct current
– Interface: Analogue

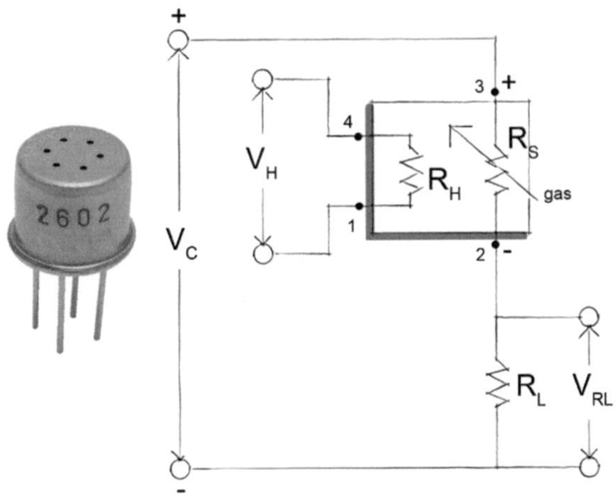

Fig. 4.18 Sensor Figaro TGS2602—TVOC measurement. The circuit scheme is re-elaborated by the producer sheet

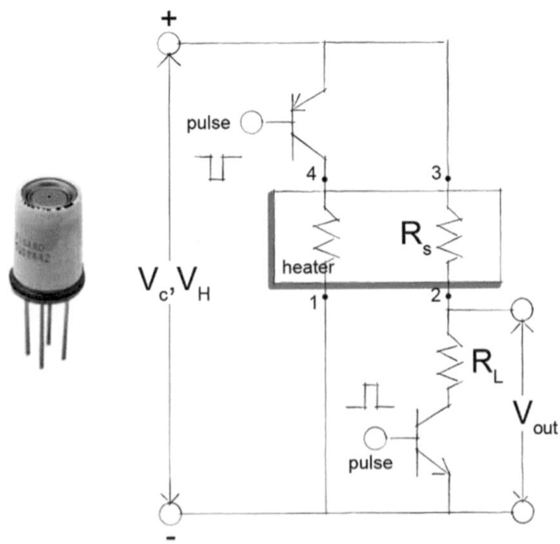

Fig. 4.19 Sensor Figaro TGS2442—carbon monoxide measurement. The circuit scheme is re-elaborated by the producer sheet

- Specifications: the sensor requires two input voltages, one to power the heating circuit to keep the temperature constant for the measurement, the other to power the measurement circuit.

Figaro TGS2442

- Gas: high sensitivity to carbon monoxide; low sensitivity to alcohol vapours
- Power supply: pulse preheating cycle at 4.8 ± 0.2 V; pulse voltage cycle at 5 ± 0.2 V
- Interface: Analogue
- Specifications: low dependence on the humidity value, it requires an electric circuit with intermittent operation with one-second heating and voltage cycles in order to read the concentration of the gas based on the conductivity of the sensor. The preheating cycle consists of a 14 ms charging time at 4.8 V followed by 986 ms at 0 V; the voltage cycle envisages 995 ms at 0 V followed by 5 ms at 5 V; the reading is taken in the middle of the charging time at 5 V (997.5 ms from the start of the cycle). The sensor requires a pre-heating period of at least two days, as illustrated in the technical specifications.
- Note: to manage the cycles it is necessary to use digital PWM pins on the Arduino boards.

Other sensors

DHT11 sensors were used to measure the RH and dry bulb temperature, DS18B20 to measure temperature only, SHARP GP2Y1010AU0F for particulates.

DHT11

- Variables measured: temperature and relative humidity
- Power supply 5 Vdc
- Humidity
- Measuring range 20–90%
- Accuracy ±5%
- Temperature
- Measuring range from 0 to 50 °C
- Accuracy ±2 °C

DS18B20

- Variable measured: temperature
- Power supply from 3 to 5.5 Vdc
- Measuring range −55 to +125 °C
- Maximum resolution 12-bit—however the Arduino UNO R3 board works with a 10-bit resolution on analogue pins
- Accuracy in the range −10 + 85 °C equal to ±0.5 °C

DFROBOT LM35 linear temperature sensor

- Variable measured: temperature
- Sensor based on National Semiconductor Corporation's LM35 semiconductor
- Power supply
- Measuring range 0–100 °C (DfRobot)
- Sensitivity 10 mV every °C
- Accuracy 0.5–25 °C

Sharp GP2Y1010AU0F

- Variable measured: particulates
- Power supply from 5 to 7 Vdc
- Operating temperature: −10 to +65 °C
- Current consumption: 20 mA maximum
- Technology: infrared diode in emission and a phototransistor positioned diagonally to record the radiation reflected by the particulate.
- Output: analogue
- Specifications: the voltage can be considered proportional to the quantity of particulate present (0.5 V every 0.1 mg/m^3) in the range 0–0.5 mg/m^3 (DfRobot). The manufacturer also supplies an indicative conversion graph, although it would be necessary to reconstruct it for each individual sensor if more precise measurements are required.
- Note: the sensor connection pins have a different pitch than the connectors generally used, so it is necessary to have a specific plug.

TSL2561 digital brightness sensor (Adafruit)

- Variable measured: brightness
- Power supply: 2.7–3.6 V
- Operating temperature: −30 to +80 °C
- Dynamic operating range: from 0.1 to 40,000 lx
- Specifications: The TSL2561 comes already assembled but it is necessary to solder a 6-pin header in order to make the connections easily. The software list is much more articulate than for the DfRobot sensor—described below—but you can use specific libraries developed by Adafruit to directly obtain the output value in lux.

DfRobot Analog ambient light sensor

- Variable measured: brightness
- Power supply: 3–5 V
- Interface: analogue
- Technology: Photoresistor, returns a voltage value (direct current) based on the amount of light that reaches the instrument—high brightness corresponds to lower values

Arduino GPS Schield v1.1 (Iteadstudio)

- Variable measured: geopositioning
- Global Positioning System—receiver and SD card slot
- Manufacturer: ITEAD Studio (also distributed by Hong Kong Worthy Electronics Co. Ltd)
- Power supply: compatible with both 3.3 and 5 V pins
- Operating temperature: −40 to +85 °C
- Sensor module: u-blox NEO-6M GPS
- Specifications: the board is compatible with Arduino MEGA, Arduino UNO and Arduino Leonardo
- Note: the complete list of other boards for which compatibility is guaranteed is supplied on the manufacturer's datasheet.

4.5.2 Sample List of Microcontroller Boards

The heart of each monitoring station is the microcontroller board, the component that manages the I/O pins, the power supply of the sensor, any expansion boards and the methods of data transmission and storage. The board also houses the software for the operation of the node.

The list of microcontroller boards that can be used for applications for the widespread production of data and are compatible with a DIY (Do It Yourself) approach is long and is getting longer all the time. In this paragraph, reference is made to some Arduino boards and those derived from the same standard. The Raspberry Pi micro-computer is also introduced, although it has fewer I/O ports and was not natively developed for permanent installations. However, Raspberry Pi (micro-computer)—Arduino (microcontroller for sensor management) integrated solutions can be envisaged, as suggested by recent studies (Dennis 2013; Monk 2010). These solutions allow easier management of the single monitoring node, which acquires new potential for the transmission, storage and direct display of values monitored through the video and audio output channels of Raspberry Pi, and of domestic actuators in cases where there is a prevalence of software. The connection between the two solutions can be a way to unite different functions in the single node, increasing its dissemination potential. It is possible to combine the functions of a monitoring station installed as part of a network and of sensor/actuator nodes for the home automation system. The integration between the two boards also allows the direct analysis of the data in the Raspberry Pi nodes. This analysis can use specialised software. A free license is available for Mathematica, developed especially for the Raspberry Pi platform and supplied by Wolfram (Mathematica's development company).

It is possible to compare the characteristics of principal Arduino Boards currently on the market by opening the following link: http://arduino.cc/en/Products.Compare (last view April 2019).

It is also possible to purchase numerous clones of the Arduino boards, some of which have been modified compared to the original with the addition of particular specifications. These include the DfRobot cards (called DFRduino) and the NETduino cards. These clones, derived from Arduino open hardware, are very cheap. A significant example is the campaign by Harold Timmis on the *Indiegogo.com* platform, to raise funds to produce an Arduino Leonardo kit for just $9+ shipping costs. The *crowdfunding* operation launched on the 12th of July 2013 was successfully concluded on the 10th of August. Asian manufacturers also allow the purchase of Arduino clone cards, developed in compliance with the open hardware philosophy, at equally competitive prices.

Arduino compatible boards integrated with processor systems

In the previous mentioned link for Arduino boards' comparison, they are also shown the characteristics of some particular Arduino boards equipped with a Linux running processor connected to a microcontroller (e.g. Arduino Yun R2). In addition to Industrial 101 and Arduino Yùn, there is also the INTEL Galileo Development board (generation 1—marketed from January 2014) with special regards to the generation 2.

The Galileo gen 2 microcontroller board is based on an Intel® Quark SoC X1000 32bi 400 MHz processor (Intel Pentium class). The board is designed to work at both 3.3 and 5 V and has a power (barrel) of 7–15 V and allows 8/12-bit of resolution on PWM pins. Galileo is also compatible with Arduino's development software. The board incorporates some additional computer-specific ports such as a mini-PCI Express slot, a 100 Mb Ethernet port, a micro SD slot, a serial port, a host USB port and a pico client port, and an 8 Mb NOR flash memory. The Galileo board is certified by Arduino and can be found in the appropriate section of *Arduino.cc*.

Last but not least is the Raspberry Pi card, a real computer made in England and marketed since 2012. The CPU of this product was initially the ARM1176JZF-S with a frequency of 700 MHz. There are several versions of board on the market, including the Raspberry Pi 3 Model B+, with a Broadcom BCM2837B0, a Cortex-A53 (ARMv8) 64-bit, 1.4 GHz, 1 Gb of SDRAM, a wireless and Bluetooth connection, a Gigabit Ethernet port, 4 USB port 2.0, and the recent Raspberry Pi 3 A+, which has the same processor, but 512 MB of SDRAM and a single USB port 2.0. Both models feature a micro SD slot, a full-hdmi video and audio output, and two 20-pin digital files. The board is powered by a 5 V current and supports Linux-based operating systems. Starting from the Rasberry Pi 2 board, the processors installed allow the installation of the highest performing operating systems, such as Snappy Ubuntu Core or Microsoft Windows 10. Unlike Arduino, the Raspberry has no analogue pins.

4.5.3 *Sample Communication, Data-Collection and Data-Storage Approaches*

IoT systems are based on specific protocols. However, in order to make the communication methods comprehensible in systems that are easy to replicate, some of the main transmission methods available are listed here.

Each station can be implemented for connection to different transmission and data collection networks. It is possible to obtain stand-alone nodes, capable of recording data on SD micro cards or on the internal memory of the microprocessor board, despite this being very limited. The EEPROM of Arduino UNO R3 is only 1 kb for example. You can also record data on your computer via USB connection, wireless cards and Bluetooth. In order to manage a network of nodes remotely, it is possible to design networks connected to the Internet capable of storing and sending data using specific servers. There are Arduino-compatible expansion boards capable of managing Xbee radio modules, which allow the construction of point-to-multipoint and mesh networks. It is also possible to use Ethernet expansion boards and Bluetooth cards. Arduino Yùn Rev 2 integrates an Ethernet and WiFi connection capable of communicating directly online using the Linux-based Operating System installed on the board for managing connections.

Boards and protocols for data transmission and communication

A microcontroller board can be connected to a network for transmitting and receiving information. There are many ways to manage embedded gateways, i.e. packet information forwarding systems, such as Bluetooth, CAN, ethernet, GPRS, HomePlug, RFID, USB, WiFi, X-10, Z-Wave connections. The main options for physically building an internet gateway however, are three: ethernet, WiFi and systems for mobile devices, such as GPRS, 2 G, 3 G, 4 G (Faludi 2011).

Some data transmission methods that are compatible with the Arduino platform are described below.

USB TRANSMISSION

Almost all Arduino boards, except some LilyPad, Arduino Pro and Arduino Mini boards, have a USB port for direct connection to a computer. This makes it possible to program the board and also send data to the computer using a COM port, depending on the settings of the software list. It is also possible to send and receive data to and from the board through applications such as Firefly.

To use the serial port it must be programmed in the program setup using the command:
Serial.begin(nr baud);

During the loops it is possible to write the values defined on the port identified using:

Serial.print("word");
Serial.println(variable);

The commands "print" and "println" are different because the suffix "ln" indicates that the next value written on the serial port is not on the same line but at the beginning of the next one.

Remember that to connect an Arduino node to the Internet via serial/USB you need to go through the connection of the computer that it is connected to.

Ethernet transmission

Specific expansion boards, such as the Arduino ethernet Shield, allow connection of an Ethernet cable to the node for connection to a network. The Arduino software includes a library dedicated to the management of the Ethernet connection to be included at the beginning of the software list with the wording:
#Include <Ethernet.h>

The card communicates with the network using the SPI bus, and the specific library (*#include <SPI.h>*). To make it work you must also assign the board an IP address, which depends on the network you are connecting to, and a MAC address. The MAC code is a globally unique ID. The latest cards have a dedicated MAC address supplied with the hardware component. If this code is not present, however, it will be necessary to assign one to the card by trial and error. You can use a dynamic IP assignment, if required, and indicate the network gateway and subnet. By way of example, here is an excerpt from a software list that uses an ethernet card:

```
byte mac[] = {0xDE, 0xAD, 0xBE, 0xEF, 0xFE, 0xBE};
byte ip[] = {192, 168, 1, 155};
byte gateway[] = {192, 168, 1, 130};
byte subnet[] = {255, 255, 255, 0};
EthernetServer Servers (80);
String readString = String (30);
```

The data transmitted by the Ethernet card can be viewed by typing in the IP of the Arduino node from a device connected to the local network or, in the case of a global connection, to the Internet. In the previous example you will write:
http://192.168.1.155/?out=all

WiFi transmission

The transmission over wi-fi network can be done using specific expansion cards. The Arduino WiFi shield, for example, allows you to connect the Arduino to the internet using the WiFi 802.11 specification (802.11b and 802.11g). The use of the specific WiFi library allows you to easily write program lists with internet connection.

Wireless transmission (ZigBee and 802.15.4 protocols)

The Arduino Wireless SD shield and the Wireless Shield allow connection to an Arduino node to a wireless network via an additional wireless module. These boards are designed to work with DIGI's Xbee modules that can replace serial connections via USB and constitute different types of networks and broadcast systems. The Xbee modules are inexpensive and allow radio transfer using a serial communication that makes them easy to use via the UART peripheral of the microcontroller or a serial port of a computer (Bernardo 2011). There are different types of Xbee board on the market. These cards are mainly distinguished by the type of communication protocol used. To find your way around the various factsheets on the market you can refer to the cataloguing developed by Bernardo (Bernardo 2011) although the offer has been partially updated. Currently, the manufacturer's website (November 2018) divides the cards into 5 areas: the 3 series—Xbee3, which includes Xbee 3 2.4 Ghz Zigbee 3.0, Xbee 3 DigiMesh, Xbee 3 802.15.4 normal version and Pro and Xbee 3 cellular—the Xbee Wi-Fi—802.11/g/n, e.g. the Xbee Wi-Fi S6B—the Xbee Zigbee—e.g. Xbee Pro S2C Zigbee, Xbee S2C Zigbee, and Xbee S2D Zigbee (thread ready module)—the Xbee 802.15.4—e.g. Xbee Pro S2C 802.15.4, Xbee S2C 802.15.4, Xbee S1 802.15.4 normal and Pro—and the Xbee DigiMesh 2.4—e.g. Xbee S2C DigiMesh 2.4 normal and Pro version, and the Xbee S1 DigiMesh 2.4 normal and Pro version.

The two series (1 and 2), to which the series 3 has recently been added, have different characteristics and functions. Series 1 models cannot communicate with Series 2 models and vice versa. PRO models can communicate with NON-PRO models in the same series, while Digimesh's Series 1 models are not always compatible with 802.15.4 Series 1 models (Bernardo 2011).

Xbee cards that use the 802.15.4 protocol allow point-to-point, point-to-multipoint and peer-to-peer communications, and this is an additional mode to the standard 802.15.4. In the first case, two devices communicate with one another, in the second case, one node is programmed as a coordinator and can receive and transmit data from and to each peripheral node while, the latter, have the ability to communicate only with the coordinator. Peer-to-peer communication, on the other hand, is multipoint communication without a coordinator.

The Digimesh protocol used by some cards allows the establishment of mesh networks in peer-to-peer mode. The ZigBee protocols, on the other hand, are mesh networks, characterised by the fact that two nodes placed too far apart to communicate directly can communicate with each other through the automatic relaunch of information by other Xbee modules in the same network. These intermediate nodes act as routers. The communication path changes automatically if the intermediate nodes connected to the network change.

In addition to the type of communication protocol and the Series, Xbee boards are distinguished by model. There are PRO or non-pro models that are distinguished mainly by the transmission power, the distance covered by the board, and the type of antenna, which can be printed, as in the case of PCB and chip, wire or external, with U.FL or RP-SMA connector). An alphanumeric code on the back of the individual Xbee components guarantees their easy identification.

The Xbee modules are connected to the computer by a special board that houses an Xbee module and connects it to the PC via serial/USB. The modules are configured using specific programs such as the XCTU utility, produced by DIGI and distributed free of charge, or Tera Term and CoolTerm software. Further details can be found in literature (Allan and Bradford 2013; Bernardo 2011; Faludi 2011). Remote management is also available for remote monitoring.

Xbee modules allow easy implementation of specially designed communication networks. The nodes can be connected to the Internet using specific modules, such as DIGI's ConnectPorts, capable of creating Xbee Internet Gateways. Xbee Wi-Fi and Digi Xbee cellular have recently added.

GSM/ GPRS transmission

Data can be transmitted using the GPRS wireless network by connecting an expansion board, such as the Arduino GSM shield, equipped with a SIM card from an operator that provides coverage of the GPRS network, to the monitoring stations. This solution also allows voice or SMS telephone communication, if a microphone and a speaker are connected to the node and a software support is envisaged.

Modes and solutions for data storage

There are numerous solutions for the storage of the data monitored. Three main solutions can be identified:

- storage on an SD micro card, integrated into the monitoring station using an expansion board or directly on the microcontroller board where possible, which allows the recording of a file—e.g. *.txt—that can be opened in Excel. Arduino supplies a specific library to simplify use of the software;
- storage on the computer's HD, using the USB as serial port. Recording can be carried out using macros for direct transfer to Excel, serial port datalogging programs, using Grasshopper's Firefly plug-in, which allows the storage of data in a Rhino environment and transmission to Excel, *.txt, * files. XML, *. KML, or using a UDP protocol based on specific plug-ins;
- storage on the Internet. There are sites developed specifically for recording information from real-time monitoring, available for payment or free of charge, such as Pachube, which has been bought by Xively. It is also possible to use FTP servers.

The systems described in this chapter are open source and open hardware and guarantee extensive implementation. It is possible to develop special monitoring nodes with more memory or connect the Arduino boards with HD shared on a local network or on the Internet. SD cards can be used to create a backup copy of the data sent or for storing data awaiting transferral. This transfer can take place at set time intervals, in manual mode or, lastly, when the node intercepts active internet connections in the vicinity.

4.6 Conclusions

The monitoring simulation presented in Sect. 4.4 demonstrated the simplicity of developing and connecting a single DIY node suitable for a widespread monitoring campaign. The data collected also reveal high sensitivity to change, suggesting air quality analysis scenarios and other indoor and outdoor variables based on low-cost widespread sensors. Similar examples are being developed by many universities and research centres around the world. Even Google is implementing innovative applications and functions based on big data analysis. These functions monitor traffic in real time, as visible in Google Earth or on the various mobile Google apps, as well as the spread of disease and infections. Many companies use big data to achieve their goals more precisely and quickly than traditional measurement methods. An example is Xoom, a money transfer company, which was able to quickly identify and stop the spread of cloned ATM cards in the state of New Jersey in 2011 (Mayer-Schönberger and Cukier 2013, 44; Rosenthal 2012).

The production, collection, analysis, optimisation, visualisation and use of data, specifically real-time big data, will radically change management and production methods and the way information is measured and analysed. Anyone will be able to produce individual monitoring nodes only using the information disseminated online. The use of big data will have to be organised for specific purposes by expert groups, building ad hoc scenarios. It is essential to have metadata on the stations and on the individual values, to allow automatic validation. In the case of bottom-up networks, such as that suggested in this chapter, it will be possible to manage this specific aspect thanks to the implementation of communities, books, blogs and kits developed specifically for makers.

It should be noted that, from the point of view of big data, the presence of individual values that deviate from expectations does not significantly change the quality of the final result, unlike what happens in traditional scientific monitoring. The extension of the sample in relation to the total population is radically different. Big data are characterised by a large number of datapoints, while traditional monitoring is based on databases containing few significant data (Mayer-Schönberger and Cukier 2013; Nielsen 2012; Shirky 2010).

Datamining and data analysis techniques are developing rapidly, following strategies linked to the concepts of collective intelligence and widespread knowledge (Janert 2011; Segarant 2007; Larose 2005). New open source and proprietary tools allow the implementation of new ways of managing relational databases and datasets (Warden 2011).

The development of innovative solutions for environmental monitoring is demonstrated by recent work in the literature (Di Justo and Gertz 2013; Gertz and Di Justo 2012) and by the dissemination of projects promoted by numerous companies, such as Libelium, and public bodies, such as the City of London, Paris, Boston, Dallas and New York—a growing list. It is presumed that this interest may increase in the future, opening up new market areas, as underlined, in several areas, including building—see for example the specific calls within the sphere of Horizon and the request to include smart elements in calls related to other areas, such as energy.

The dissemination of information and data will also make it possible in future to strengthen development and research in the field of parametric urban planning. The Singapore City Form Lab is developing application scenarios based on the effects that big data are having on the optimisation of the planning of services and activities. Within the scope of big data research, it will be necessary to develop innovative ways based on the "*what*" and not only on the "*why*", creating new scenarios for the construction of causal links (Mayer-Schönberger and Cukier 2013).

Acknowledgements In order to develop this chapter, voluntary use of the knowledge available online was made. A passive questioning of the knowledge disseminated was chosen in order to verify its usefulness. We must thank those who helped build the information present online. Thanks to Pierangelo Stradella for his support with the construction of the ethernet station presented in the chapter.

References

Acquaviva A, Apiletti D, Attanasio A, Baralisi E, Bottaccioli L, Castagnetti FB, Cerquitelli T, Chiusano S, Macii E, Martellacci D, Patti E (2015) Energy signature analysis:knowledge at your fingertips. In: Proceedings of the IEEE International Congress on Big Data (BigData Congress), New York, 27 June–2 July, pp 543–550

Allan A, Bradford K (2013) Distributed network data. O'Reilly, Sebastopol

Anderson C (2012) Makers: the new industrial revolution. Random House Business Books, London

Benammar M, Abdaoui A, Ahmad SHM, Touati F, Kadri A (2018) A modular iot platform for real-time indoor air quality monitoring. Sensors 18:18p

Bernardo G (2011) Guida alla scelta e alla comprensione dei moduli XBee—Rev.1, distribuito gratuitamente da www.robot-italy.com e da www.settorezero.com, e.g. http://www.robot-italy.com/download/xbee/easy_bee.pdf. Last view Apr 2019

Chiesa G (2010) Biomimetica, tecnologia e innovazione per l'architettura. Celid, Torino

Chiesa G (2013a) La città digitale, dai sensori ai modelli: Piattaforme interconnesse per la città del futuro. In: Di Giulio R et al (eds) Strategie di riqualificazione urbana: Rigenerazione e valorizzazione dell'edilizia sociale ad alta densità abitativa del secondo Novecento. Quodlibet, Macerata, pp 110–117

Chiesa G (2013b) M.E.T.R.O. (Monitoring energy and technological real time data for optimization) innovative responsive conception for city futures. Ph.D. Thesis, Politecnico di Torino, Torino

Chiesa G (2014) Data, BigData and smart cities. Considerations and case study on environmental monitoring. Techne 08:81–89

Chiesa G (2017) Explicit-digital design practice and possible areas of implication. Techne 13:236–242

Chiesa G, La Riccia L (2013) Dalla rappresentazione alle rappresentazioni di paesaggi e territori. Planum 27(2)

Chiesa G, Palme M (2018) Assessing climate change and urban heat island vulnerabilities in a built environment. Techne 15:237–245

Chiesa G, Grosso M, Acquaviva A, Makhlouf B, Tumiatti A (2018) Insulation, building mass and airflows—provisional and multi-variable analysis. SMC-Sustainable Mediterranean Construction 8:36–40

Chiesa G, Acquaviva A, Grosso M, Bottaccioli L, Floridia M, Pristeri E, Sanna EM (2019) Parametric optimization of window-to-wall ratio for passive buildings adopting a scripting methodology to dynamic-energy simulation. Sustainability 11:30. https://doi.org/10.3390/su11113078

Clarke JA (2011) Excellence Ph.D. course simulation-based CAD: delivering detailed building performance information in real time. Politecnico di Torino, Turin, Italy

Dennis AK (2013) Raspberry Pi home automation with Arduino. Pack publishing, Birmingham

Di Justo P, Gertz E (2013) Atmospheric monitoring with Arduino. O'Reilly, Sebastopol

Faludi R (2011) Building wireless sensor networks. O'Reilly, Sebastopol

Ferraris M (2003) Ontologia e oggetti sociali. In: Floridi L (ed) Linee di Ricerca. SWIF, pp 269–309. Viewed Jan 2014. www.swif.it/biblioteca/lr

Gertz E, Di Justo P (2012) Environmental monitoring with Arduino. O'Reilly, Sebastopol

Ginsberg J et al (2009) Detecting influenza epidemics using search engine query data. Nature 457:1012–1015

Girardin F et al. (2007) Understanding of tourist dynamics from explicitly disclosed location information. In: 4th international symposium on LBS and telecartography. Hong-Kong, China

Givoni B (1998) Climate considerations in building and urban design. Wiley, Van Nostrand Reinhold

Hashimoto S, Dijkstra R (2004) Chip city. In: Ferré A et al (eds) Verb Connection, Architecture Boogazine. Actar, Barcelona, pp 46–53

Heiselberg P (ed) (2018) Ventilative cooling design guide. IEA EBC Annex 62. Aalborg University Press, Aalborg

http://www.indiegogo.com/projects/9-arduino-compatible-starter-kit-anyone-can-learn-electronics. Last view Apr 2019

http://www.dfrobot.com. Last view Apr 2019

http://arduino.cc/. Last view Apr 2019

http://blog.iteadstudio.com/. Last view Apr 2019

http://www.wolfram.com/raspberry-pi/. Last view Apr 2019

http://www.digi.com. Last view Apr 2019

http://www.comune.torino.it/geoportale/. Last view Apr 2019

http://senseable.mit.edu/. Last view Apr 2019

http://cityform.mit.edu/. Last view Apr 2019

http://netduino.com/. Last view Jan 2014

http://ec.europa.eu/transport/its/index__en.htm. Last view May 2010

https://www.covenantofmayors.eu/. Last view Nov 2018

http://www.robot-domestici.com/. Last view Oct 2013

Janert PK (2011) Data analysis with open source tools. O'Reilly, Sebastopol

Khabazi M (2009) Algorithmic modelling with grasshopper (Rhino Plug-in). Last view Dec 2010. www.khabazi.com/flux [no longer available], now accessible at https://www.pinterest.it/pin/191543790375951325/. Last view Apr 2019

Larose DT (2005) Discovering knowledge in data. An introduction to data mining. Wiley, Hoboken

Libelium (2012) GASES 2.0 Technical Guide, technical report project Waspmote ver. 2-0.2, 11/2012, Libelium Comunicaciones Distribuidas S.L., last view 2013. http://www.libelium.com/waspmote

Libelium (2015) Agriculture 2.0. Technical guide. Waspmote, v.5.0. Libelium Comunicaciones Distribuidas S.L., Zaragoza

Libelium (2016) Events 2.0. Technical guide. Waspmote, v.5.1. Libelium Comunicaciones Distribuidas S.L., Zaragoza

Libelium (2018a) Waspmote Plug&Sense! Technical Guide, v8.0. Libelium Comunicaciones Distribuidas S.L., Zaragoza

Libelium (2018b) Smart Cities Pro. Technical guide. Waspmote, v7.5. Libelium Comunicaciones Distribuidas S.L., Zaragoza

Libelium (2018c) Smart Gases Pro. Technical guide. Waspmote, v7.5. Libelium Comunicaciones Distribuidas S.L., Zaragoza

Mayer-Schönberger V, Cukier K (2013) Big data. Una rivoluzione che trasformerà il nostro modo di vivere e già minaccia la nostra libertà. Garzanti, Milano (ed. or. (2013) Big data: a revolution that will transform how we live, work, and think. Houghton Mifflin Harcourt, Boston)

Monk S (2010) 30 Arduino projects for the evil genius. Tab Books, Blue Ridge Summit

Myrup LO, Morgan DL (1972) Numerical model of the urban atmosphere: The city-surface interface (vol I). Department of Agricultural Engineering, Department of Water Science and Engineering, University of California, Davis, CA, Oct 1972

Neumann G, Noda T, Kawaoka Y (2009) Emergence and pandemic potential of swine-origin H1N1 influenza virus. Nature 459:931–939

Nielsen M (2012) Le nuove vie della scoperta scientifica. Come l'intelligenza collettiva sta cambiando la scienza. Einaudi, Torino (or. ed. (2012) Reinventing discovery: the new era of networked science. Princeton University Press, Princeton)

Nowak DJ (1994) Air pollution removal by Chicago's urban forest. In: McPherson EG et al. (eds) Chicago's urban forest ecosystem. Results of the Chicago urban forest climate project. Forest Service, US Department of Agriculture, NE, pp 63–81

Odifreddi (1994) Epistemologia e Ontologia virtuali. In: Cenacolo Interdipartimentale di Torino, 27 May 1994, Torino, Italy

Osello A, Acquaviva A, Del Giudice M, Patti E, Rapetti N (2016) District information models. The DIMMER project: BIM tools for the urban scale. In: Pagani R, Chiesa G (eds) *Urban data. Tools and methods towards the algorithmic city.* FrancoAngeli, Milano, pp 231–261

Payne A (2011) Interactive Prototyping. An introduction to physical computing using Arduino, Grasshopper, and Firefly. Last view Apr 2019. http://fireflyexperiments.com/resources/

Premeaux E, Evans B (2011) Arduino projects to save the world. Apress, Springer Science distribution, New York

Richardshon M, Wallace S (2013) Getting started with Raspberry Pi. O'Reilly, Sebastopol

Riley M (2012) Programming your home. Automate with Arduino, Android, and your computer. The Pragmatic Programmers, Dallas & Raleigh

Rivoltella PC (ed) (2010) Ontologia della comunicazione educative. Metodo, ricerca, formazione. Vita e pensiero, Milano

Rosenthal J (2012) Special report: international banking. Big data crunching the numbers. The Economist, 19 May 2012, pp 7–8. Last view Apr 2019. http://www.economist.com/node/21554743 and http://media.economist.com/sites/default/files/sponsorships/MM152/20120519_international_banking_HSBC.pdf

Santamouris M (ed) (2001) Energy and climate in the urban built environment. James and James, London

Santamouris M (2011) Heat Island research in Europe: The state of the art. Adv Build Energy Res 1(1):123–150

Santamouris M, Haddad S, Fiorito F, Osmond P, Ding L, Prasad DK, Zhai X (2017) Urban heat island and overheating characteristics in Sydney, Australia. An analysis of multiyear measurements. *Sustainability*, 9(5): 21 p

Segarant T (2007) Programming collective intelligence. O'Reilly, Sebastopol

Shirky C (2010) Surplus Cognitivo. Creatività e generosità nell'era digitale. Codice Edizioni, Torino (or. ed. (2010) Cognitive surplus: creativity and generosity in a connected age. Penguin Group, London)

Vico F, Lucà R (2005) DATASCAPES urbani. Conference ASITA 9° Conferenza Nazionale, Catania 15–18 Nov 2005

Wang S (2010) Intelligent buildings and building automation. Spon Press, London

Warden P (2011) Big data glossary. O'Reilly, Sebastopol

Weinberger D (2012) La stanza intelligente. La conoscenza come proprietà della rete. Codice edizioni, Torino (or. ed. (2011) Too big to know: rethinking knowledge now that the facts aren't the facts, experts are everywhere, and the smartness person in the room is the room. Basic Book, New York)

Yang J, Santamouris M (2018) Editorial. Urban heat island and mitigation technologies in Asian and Australian Cities—impact and mitigation. *Urban Science* 2:6 p

Chapter 5
Platform—A Space for Project Design and an Interface Between Reality and Virtuality

5.1 The Platform Concept—A Space to Manage and Design Complexity

The first and second digital ages are articulated around the concepts of virtualisation of reality and materialisation of virtuality, with a view to transforming from implicit to explicit links between the different components and knowledge of design and implementation that contribute to the project. These two opposite processes are applied in the concept of platform, a place of connection between reality and virtuality (Chiesa 2010, 2015; Pagani 2009).

The term platform refers to "a flat surface obtained artificially by levelling and hammering out a stretch of land, or consisting of a board raised more or less from the ground or from other structures [but it may have other meanings, including] machine tool element with rotary cutting motion, used for assembly the piece with jaws or bolts [and] in information technology, the set of hardware technology and operating system that characterise a computer" (Devoto and Oli 2009). In short, the platform is a base, either software or hardware, on which to organise the design, construction and use intended as the execution of an application.

The platform is also the intersection space between flows, inhabitants, technologies and software, hardware and orgware services on different scales (local network, regional network, global network) from the viewpoint of the global city (Chiesa 2010; Pagani 2009). The physical/virtual space of the platform is an evolutionary place, which can be customised to suit the occupants and their needs on the basis of a series of specific parameters such as connections to electricity, gas, internet, telephone, TV, shops, but also information, databases, access codes and knowledge (Fig. 5.1).

It is a flexible space, adaptable, modifiable and implementable in time and space, whether virtual or real. The platform can also be configured as a tool for managing models and information, as well as the sphere for the design of complexity. In turn, a platform can present ramified sub-platforms called plug-ins, which, thanks to specific tools and services, can provide different ways of modelling and materialising data

G. Chiesa, *Technological Paradigms and Digital Eras*,
PoliTO Springer Series, https://doi.org/10.1007/978-3-030-26199-3_5

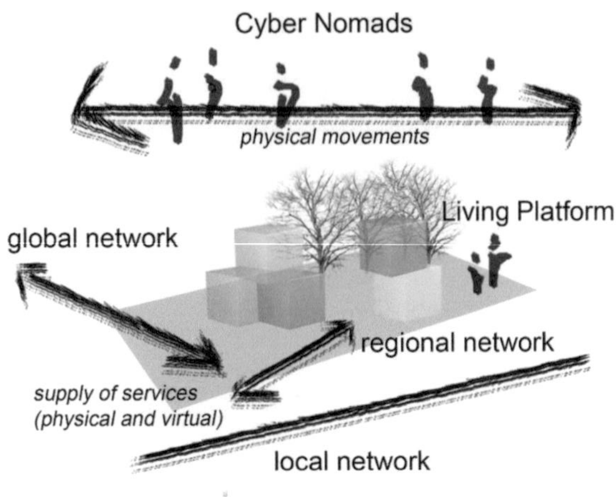

Fig. 5.1 Functional schematic representation of a flexible platform conceived to perform users' needs related to physical movements of people in the global world. Re-elaboration of the "cyber nomad" idea representing a network of nodes for new nomads in the global era—see Chiesa (2017, 2010, 113), Pagani (2009)

and geometry. It can be a place for the development of solutions, a place for the dissemination of results, a space for experimentation.

Changing the sphere of application, the platform is configured as the place of management of complex designs and smartness, on different scales. It can be both node and network because it represents the space of the process, of the conjunction and of the creation of models and information. Data, information and process methods for the construction of models which, as Maldonado reminds us, have been intended as design tools since the birth of the modern figure of the architect, converge within the platform. In short, the model represents the analogue or digital object in which the passage from reality to virtuality takes place and, at the same time, is the instrument through which virtuality becomes reality. The transformation space is the platform that governs the processes of modelling and materialisation. The latter process can take place in different geographical locations to the process of creating the model for materialisation, spreading the production spaces across the territory. This concept is the basis of the new industrial revolution (Anderson 2012), defined in this text as the second digital age.

Alongside this industrial innovation, a process of virtualisation of the city connected with the massive production and use of data, i.e. the third digital age, is taking place. Smart cities can be considered as platforms for the interconnection of real spaces, data production (sensors) and actuation (materialisation by actuators), while virtual spaces for processing, interpretation and decision in the sense of virtual, hybrid, semi-hybrid spaces can be considered as datascapes and model spaces. However, it is also possible to produce and "materialise" data and information in virtual spaces. Within Smart Cities, one or more local platforms can be identified for each city to help manage the specific vision of the future, bring together stakeholders, prepare calls for funding and organise the local and global resources and technologies available. At the same time, platforms on different geographical scales are configured. As far as the smaller scales are concerned, these are regional, national and supranational platforms, brought about by the movement of goods, people and international relations on different scales, by the implications of rules and choices dictated by the various parameters capable of influencing the forms of management and visualisation/materialisation of space. In the case of platforms on a large geographical scale, the focus is on districts, blocks, buildings and apartments, in terms of integration and organisational capacity necessary for the development of a real urban resilience and to achieve thresholds of energy-environmental sustainability typical of the net positive future in compliance with a mature perspective.

An important value of the smart city lies in the use of data from the real world, collected in real time for popular models through which it is possible to optimise different urban functions. There can be a considerable difference between a real monitored data item, even in cases where low cost sensors are used, and a virtual data item generated through simulations. The use of large amounts of real-time data make it possible, at least in part, to shift the virtualisation and use of scientific models to the processing phase, enriching the quality and speed of the processes and ensuring better compliance with the real world, also in the simulation phase. The use of monitored data makes it possible to assess and process data and information with a view to operational rating rather than asset rating.

Data are produced and collected, also through datamining and cloud computing techniques, visualised, processed, re-visualised and used also to optimise specific functions—e.g. management of energy flows at district level. In cases where the data is derived from monitoring, the process starts from the real world, is transferred to the virtual world via the collection network and is returned to the real world in the utilisation phase. In this way, the digital spaces of representation and visualisation become interfaces between the virtual and real worlds, managed in the platforms. It is therefore possible to gradually optimise the decision-making models used through machine learning techniques that optimise the response of the actuators to changing boundary and input conditions. In this sense, platforms can be developed on the basis of the application of hardware approaches in the loop, where simulation is connected to the real world—input/output. See, for example, research into forecasting models on the productivity of photovoltaic panels in urban areas (e.g. Aliberti et al. 2018) or studies on the management of district heating flows to the final nodes in a prosumer vision.

However, the data produced and collected in an architectural and urban context, for example, may not all be the same, being either proprietary or public, conformant or otherwise to a standard that guarantees interoperability, and may differ in terms of application. Data, understood as the simple entity that will populate complex models, is the mirror of the platform that produces and collects it and, as such, is also an indicator of the type of smart city chosen by the local area. Public data may be distributed to third parties free of charge or kept confidential. ARPA, for example, supplies some of the data collected on urban air pollutants free of charge, while ISTAT supplies some kinds of data, while sensitive data, including data on justice or income, are not accessible to everyone. Some public administrations, in the wake of the project for the creation and dissemination of open data promoted by the Obama administration (Mayer-Schönberger and Cukier 2013; Weinberger 2012), are promoting similar initiatives by adding new datasets to the data that are already available. See, for example, the case of the Municipality of Turin, which has flanked the data available on the municipal geoportal with new information, including that reported on the regional open data website (www.dati.piemonte.it). Private data can be open, if they are released free of charge, in the perspective of "give away the bits and sell the atoms" (Anderson 2012), sold rough or processed to third parties for a fee, or confidential, with data on industrial secrets, tests and patents not being released. The smart city, understood as the city of data, is public or private, open or confidential, free or subject to charge. If daily services and actions interface in natural/artificial and virtual/real hybrids, and if data are at the base of the construction of space, when they become subject to charge or are proprietary and secreted, the ICT that use them show dangerous drifts based on the disabling power of IT. Mechanisms of exclusion of groups of the population are triggered along with the birth of disconnected sub-cities or those connected to other networks. The cloud and the abundance of data produced at pervasive level on the territory in different ways ranging from the concept of ownership to accessibility, reflect the Long tail concept introduced earlier. The Internet of things, the smart city and current technologies predict that the Long tail of data is only in the initial phase, as the network connection of nodes capable of collecting, communicating and transmitting data, is becoming a tangible reality for the cities of the future. As mentioned with regard to data, platforms can also be developed according to different approaches: proprietary, open access or public. Proprietary platforms, such as the one developed by *Reply* for *Internet 3.0*, are high quality tools which do, however, entail certain risks with regard to the ownership of and access to data, especially in a context where the public and private sectors come together. There are other companies that offer solutions for smart cities and open hardware, based on the same philosophy as example Arduino and Libelium Comunicaciones Distribuidas S.L. for example.

In the panorama of platforms that are being developed for the creation and management of big data, it is also possible to identify market niches populated by makers and based on the DIY concept. There are many open platforms on the market, for implementation in both software and hardware, as in the case of rapid electronic prototyping platforms for connecting low cost sensors to custom datalogging and remote data transmission tools. Examples are the Arduino platform and Raspberry Pi, each of which provides boards for the implementation of functions, DIY kits for makers and professionals, implementable and free schemes and software. Although they cannot be organised or planned, individual user applications can be developed considerably and, if brought together in digital communities, they can constitute a source of high-quality public data based on the Internet of Things. The compatibility between a bottom-up strategy and top-down but implementable approaches, such as those offered by Libelium and its competitors, tends to indicate that we are facing interoperable platforms for the creation of data.

The construction of software interfaces to connect data to modelling platforms, such as those used by architects and urban planners, shows the high degree of complexity at stake. At this level too there are open and proprietary platforms (CAD software, parametrics, BIM and GIS), interoperability issues and the need to manage the growing flow of information at every stage of the process. These analyses show that the research work, based on the concepts of production, transmission and use of data as parameters capable of influencing civilisation, is based on tangible foundations whose applications are currently available also to non-expert users. Moreover, the data resulting from different kinds of monitoring can affect design in a perspective of real-world model interaction aimed at knowledge and optimisation.

The task of managing complexity, risks and problems, and conceiving the process, lies with the design and management of the platform, project site, data interconnection, information, modelling and materialisation techniques. However, it will be essential to organise the platforms in order to maintain the level of decisions without delegating it to machines and software.

In the urban and architectural sphere it will also be essential to be able to create the right interface between the different types of knowledge involved, to avoid the extinction of designers or, on the contrary, the underestimation of the results of ICT or of the different specific skills (thermodynamics, fluid dynamics, …).

5.2 Software, Architects, Instrumental Platforms and Digital Design Issues

An interesting example of a platform, which is becoming popular in the construction sector, consists of new software for digital design. Unlike traditional computer aided design software, contemporary digital systems allow a synergistic approach to design based on multi-software platforms. Traditional CAD software products are designed to replace the paper-based design approach and behave like single programs. Now, however, we are seeing the spread of software platforms capable of performing different functions, integrating large amounts of data, connecting libraries and databases, managing specific sub-platforms with a view to integrating knowledge, such as that partially obtainable today through BIM. These real systemic design platforms can be used to manage the model/design on different scales and in different phases of its life, allowing the integration of different types of knowledge (compositional, technological, structural, energy, …), design constraints (regulations, costs, technological constraints, …), requirements and tools. The new software allows the implementation of the innovations and design methods of the first and second digital age, integrating the explicit functions to represent, generate, assess and manage performance and produce components. The platform allows the direct introduction of the results and optimisations resulting from analysis and simulations into the digital model. In the same way, geometric models are full of information derived from databases, as in GIS systems, with a view to a single design. Virtual-real interaction, and modelling and materialisation processes, can be managed in the spaces of digital software platforms. In the case of GIS, for example, map rendering helps meet specific requirements but is reductive compared to the completeness of the result of analyses, integrated in the files that make up the shapes, or multi-capacitive analysis models. Shape processing is the real goal, as it allows new digital modelling and design software to fill a model with different types of information (not just geometric) through multiple databases. BIM (Building Information Modelling), for example, is an activity, the BIM (building information model) is an object, and BIMS (Beyond Information Models) is a system. The new platforms, be they BIM, GIS or CAD object oriented, algorithmic or parametric, are configured as tools for the first and second digital age but, above all, they introduce the tools for the designer of the third age. These are real platforms, spaces for interaction, design, development and management of information flows and generation of different kinds, between the real and the virtual.

Figure 5.2 illustrates the platform and related sub-platforms of the Autodesk (February 2014). Figure 5.3 shows the platforms of the *3D Dassault Systèmes* and that of *Ghery Technologies*, an independent platform from CATIA (February 2014). Although different software products refer to the same platform, there is not always a real interaction between geometric models and other information from program to program, creating different limitations and problems (Osello et al. 2013, 2016). Moreover, the platforms described above are closed, proprietary and difficult for individual users or developers to implement, limiting their use to the main, standard applications. This choice could favour the dissemination of tools that allow action within the perspective of innovative design of the different digital ages, while allowing partial uses or uses deriving from paper-based approaches, but it could also risk reducing the potential of digital design innovation, limiting the explicit action of design processes in the wake of the development of predefined tools.

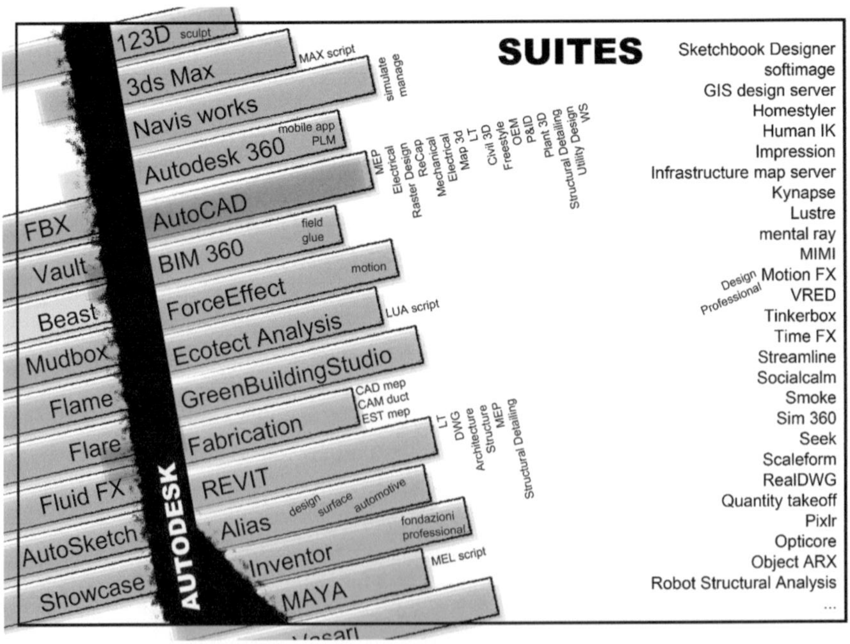

Fig. 5.2 Schematic representation of part of the Autodesk platform (February 2014)

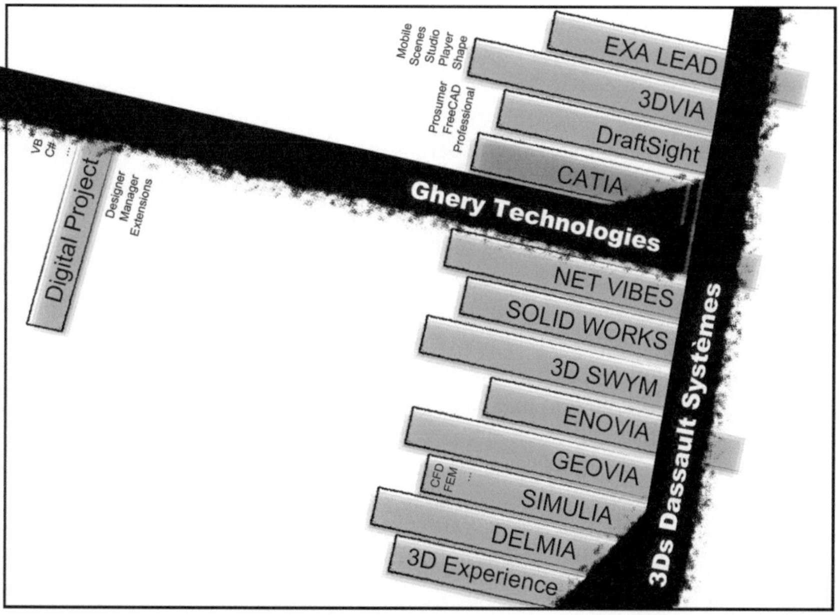

Fig. 5.3 Schematic representation of part of the 3D Dassault Systèmes platform including Ghery Technologies platform (February 2014)

For this reason, some major international architectural firms have decided to develop and design their own platforms to suit their specific needs. This decision, shared by Ghery and Foster, especially with regard to technologies of materialisation on a component and building scale, testifies to the extent of the change taking place and also indicates the need to overcome the limitations imposed by closed platforms. The latter, while being interoperable, are not flexible enough and are unsuited to the application of all the processes and interactions required, whether they belong to the first, second or third age. The cloud of most current BIM systems doesn't allow the easy integration of cloud computing techniques, being based mainly on shared work processes and sharing, as happens with the Google Drive platform, although this claim could be disproved in the near future.

A different approach is offered by open or semi-open platforms, capable of feeding and facilitating scripting and the customisation of design, maintenance and interaction methods. In this case, it is essential to remember that these platforms, in addition to performing primary functions, are constituted as nodes of interoperability between different plug-ins and sub-platforms, or allow scripting in languages such as Python (e.g. Tibbits et al. 2011) or C# (e.g. Bochicchio et al. 2010). *McNeel & Associates*' Rhinoceros-based platform, for example, allows parametric management of geometric, mathematical, physical and other data, including energy, physical, structural and economic simulations, assessments, $CO_{2\text{-eq}}$ emissions, embedded energy, LCA, big data from social networks, geographic data from Google and other platforms, various types of online connection, and database management. These data and this information can be processed in modelling processes such as geometric analysis, physical–technical analysis in dynamic mode and geometry optimisation, and in materialising processes including CAM, 3D rapid prototyping, mass customisation, making data accessible and connected between the physical and real worlds. The possibility of interaction between different specific processing software products, which need not necessarily be offered by the same IT group, allows the performance of further simulations and optimisation (Fig. 5.4—February 2014).

The CAD, BIM, GIS and CAM platforms act as an interface between different software products and are capable of being an interoperability management tool, bringing together the various potentials of different types of system. This complexity does have certain limitations, however, including the fact that it requires a good command of scripting techniques and programming languages, especially for specific applications, large amounts of memory and computing power, updates, especially for the repair of bugs that may arise in free software, and the need to redesign scripting that may become obsolete or outdated by new design tools. The potential of these platforms is, nevertheless, very interesting. Think, for example, of the plug-in for Rhino Urban Microclimate developed within MIT and capable of combining the Urban Heat Island (UWG) simulator with complex urban geometries and energy simulations—UMI software—to study part of the urban metabolism and identify

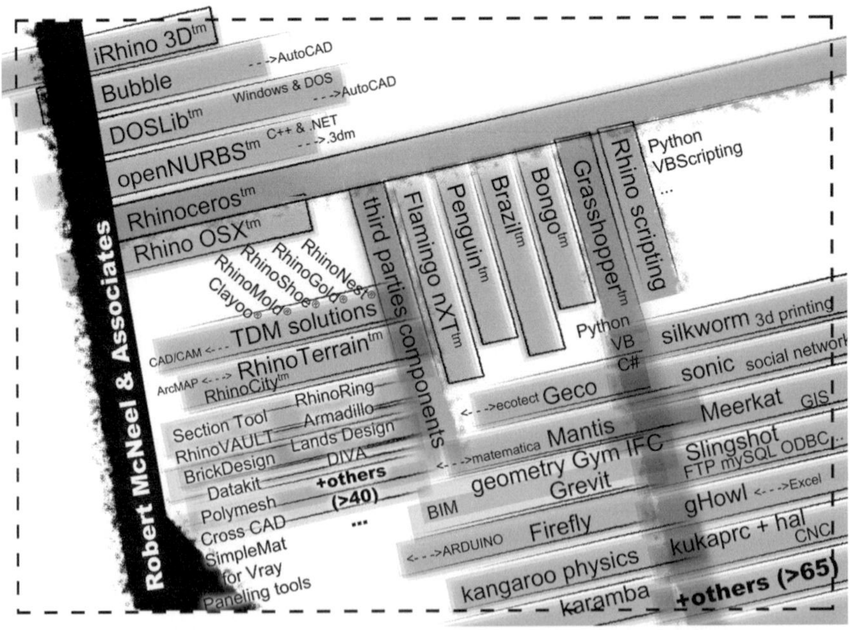

Fig. 5.4 Schematic representation of part of the Robert McNeel & Associates platform (February 2014)—sample of partially open platform. It is possible to underline the large amount of plug-ins able in increasing specific functionalities, connecting the platform with further software and other platform in a interoperable and cloud-based dimension

critical issues and strategies to solve them (e.g. Nakano et al. 2015; Nakano 2015)—see also Palme et al. (2017).

5.2.1 *Platform for Control Systems in a Reality-Virtuality-Reality Dynamic Information Flow*

The platform concept can find many applications: simulation tools, such as software for dynamic energy simulation, can be considered as design platforms. It is also now possible to predict an overcoming of the current boundaries between platforms, foreseeing an integration between geometry construction and simulation—see also Chap. 4—in order to optimise specific geometric, technological and functional choices, etc. As regards the topic of energy simulations, for example, in the Rhinoceros® platform, through Grasshopper, it is possible to use the Honeybee plug-in which, being similar to OpenStudio®, is able to launch EnergyPlus™, which is an open-source building energy simulation program funded by the U.S. DOE BTO. EnergyPlus is very diffused and used world-wide by professionals and researchers—see https://energyplus.net/ (viewed December 2018). OpenStudio is a cross-platform supporting EnergyPlus building energy modelling and daylight studying by using Radiance. Furthermore, Honeybee is part of the Ladybug tools for Grasshopper diffused under the GPL-3.0+ license. The data flow between these tools allows verification of the impact of the different design choices on the anticipated energy requirements of the buildings designed. Nevertheless, further EnergyPlus interfaces are present, such as DesignBuilder, which is a very diffused commercial tool. Although, several other potential applications are possible e.g. by coupling different software environments using Grasshopper and Python to generate a dataflow interface. For example, Perini et al. (2017) combined ENVI-Met, for its CFD potentiality, with TRNSYS using Grasshopper as data interface.

Thanks to the new scenarios enabled by IT, it is possible to foresee a new approach to the use of such simulation tools, acting directly on the input files of energy-dynamic simulation programs through scripting, in order to generate forecast models that link certain boundary conditions or technological choices with the anticipated energy consumption—see also the works (Chiesa et al. 2017a, 2019b; Huberman et al. 2015) in which the use of massive simulations to optimise shapes or calculate the applicability of certain technologies to varying locations has been managed by modifying the inputs of EnergyPlus using MathLab.

Furthermore, as part of the ICT in building design course, Politecnico di Torino, Master Degree ICT4SS, AA 2017–18, Professors Chiesa, Acquaviva and Grosso, a Python scripting system that can read EnergyPlus input files—extension *.idf—and automatically modify them by acting on specific variables has been developed. This approach allows operation on strictly design-related variables, such as the U-value (e.g. thickness of insulation), the window-to-wall ratio, or the Internal Heat

Capacity, and control of the various systems, such as controlled natural ventilation or sun screens. This Python script, also based on the Python Eppy function, allows the automatic generation of a chosen number of new *.idf by varying the values of specific parameters in defined ranges. Subsequently, these input files are processed via energyplus for different locations, generating a mass of data on anticipated energy requirements. This database has been processed to generate polynomial correlation models (e.g. from the first to the 10th degree) between the variation of the parameters considered and the anticipated energy requirements. Specific conditions can be expressed in the number of hours of discomfort, as in the case of controlled natural ventilation in residential buildings, or in the absence of a mechanical air-conditioning system. Moreover, again using a Python script, it is possible to generate random disturbance phenomena by acting on the boundary conditions of the simulations, such as the variation of thermal warnings—e.g. of the internal gains (occupancy scheduling). These phenomena are useful to statistically test the polynomial models previously developed and to identify the optimal degree of regression in order to correlate a phenomenon with the anticipated requirements even in non-sample conditions. In this sense, see the first results published in (Grosso et al. 2019; Chiesa et al. 2018). By way of example, Fig. 5.5. shows the points obtained from dynamic energy simulations without disturbance phenomena, the polynomial chosen on the basis of the statistical correspondence of these models in the presence of disturbance tests, and the datapoints of the energy requirements of the test cases with the change in occupancy rate according to a Gaussian logic applied randomly.

This approach can be adapted to specific parametric analyses of the applicability of a specific technology with a change in context and design choices (e.g. Chiesa 2019a;

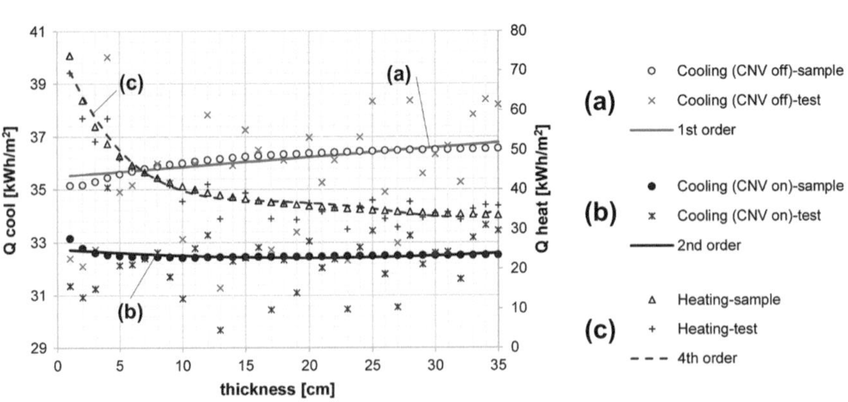

Fig. 5.5 Python coding applied to dynamic energy simulations to suggest optimal design configurations from early-design phases. In particular, the sample focuses on wall-thickness insulation levels. In this graph are shown: the sample simulated database (office case study) represented by the dotted or triangular points, the best correlation polynomial curve represented by lines, and the testing database generated on random occupancy values and represented by cross points. For this case double glazing with Argon was assumed considering cooling energy needs with and without CNV (controlled natural ventilation) and the heating energy need—see Chiesa et al. (2018)

Kosir et al. 2018; Chiesa et al. 2017b), and has enormous additional potentialities. These analyses are open and quickly adaptable to different buildings, parameters and indicators, such as was recently demonstrated for the WWR (window-to-wall ration) optimisation in different climates (Chiesa et al. 2019a) and the definition of a new process to calculate the minimal and maximal U-vales (thermal insulation) in different locations for regulatory purposes including both heating and cooling expected energy needs (e.g. Chiesa 2019b).

Another step in this innovative methodology was subsequently developed and pre-tested as part of the Interdisciplinary Project course of the same University (professors Chiesa, Grosso, Acquaviva and Dovis) to connect the results obtained from the simulations with a real cell test in order to optimise the operation of a controlled ventilation system (control of air flow rates) for space cooling purposes. This survey, carried out with the student team Tumiatti and Makhlouf, showed how it is possible to progress in this direction using the dynamic energy simulation model to optimise the ac/h (air exchange per hour) necessary to reach a given temperature as the internal gains and external temperature vary. This result is expected to be published in future while the subject is the topic of a series of forthcoming publications including machine learning approaches.

This approach can also be easily used in machine learning systems to populate neural networks in order to forecast the anticipated energy requirements when, under certain boundary conditions, a specific technological choice is made in relation to design—e.g. insulation level—and control—e.g. activation of controlled natural ventilation, control of sunscreens. This forecasting analysis can be easily enhanced during the building operation phase by entering real data, trying to optimise the feasibility of the simulations launched by automatically modifying the specific variables in the input files of software such as energyplus, and linking the results with that monitored—at least in the early stages, when the real data are insufficient to correctly estimate the operation, or in case of unforeseen stress (climate, internal gains, etc.). These data can also be used to generate an estimate model suitable for suggesting the best strategies for reducing energy consumption to users and home automation systems while ensuring the required comfort conditions. This approach can easily allow the correlation of the energy signature and energy labelling between the different phases of the building (asset and operational) and to include benefits due to smart technologies in these approaches.

5.3 Rethinking the First and Second Digital Ages Using the Platform Concept

The platform can be characterised by an instrumental value, understood as a union of tools and space of interface between man and machine, and acquire further value in terms of description of processes and innovative design methods. The conceptual schemes described in Chap. 2 for the first digital age can be translated into the concept of a platform for digital design, i.e. the space in which the different implications of this change take place and are made possible. The schematic reading of the design process on the basis of the platform concept assumes particular value in cases of integration between modelling and materialisation processes, as in the second digital age/paradigm. In this context, the platform can be defined as a modelling space for the digital design process, in which the flow of the process itself is crossed by the conjunction of four or more axes. The diagram in Fig. 5.6 illustrates this principle from the point of view of an integrated approach based on a vision of instruments, knowledge, techniques, restrictions and technologies. Each axis refers to one or more specific areas, as highlighted in the diagram, which can generally be defined as the axis of constraints, needs, requirements, laws and regulations, tools, media, materials and technologies, experiences and technical skills. In order to meet the needs of the different design levels, it is possible to define specific parameters that can be optimised in a repetitive process capable of interfacing with both the model and the flow of information. Mass customisation techniques are an important tool to satisfy this scheme, in the same way that they are for the context of digital design. Of course this approach is part of the work in process connected to design issues in where subjects, expertise, design features and design steps are all interconnected—see the scheme of Fig. 5.7 elaborated with Casetta (2009).

This scheme of Fig. 5.6 can be implemented as illustrated in Fig. 5.8, according to an approach based on the whole process, starting from the first design ideas (early-stage-design) and ending with the construction of the building. The same process can be further extended to consider the management, maintenance, end of life, decommissioning and, where applicable, recovery phases.

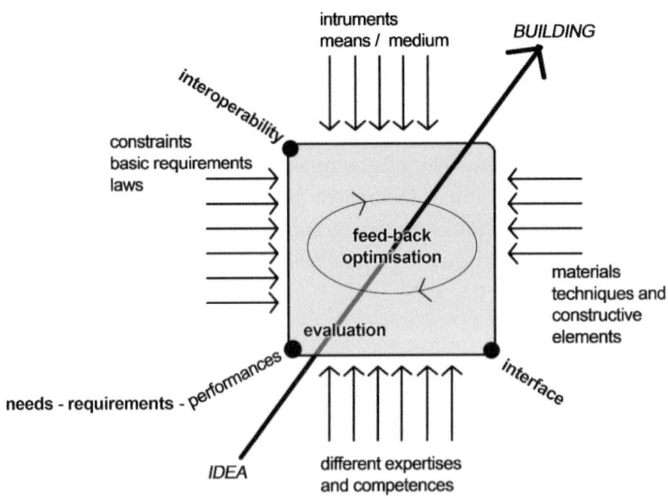

Fig. 5.6 General schematic description of the multi-axes platform for a multi-systemic and technological-integrated vision of the design process in digital eras

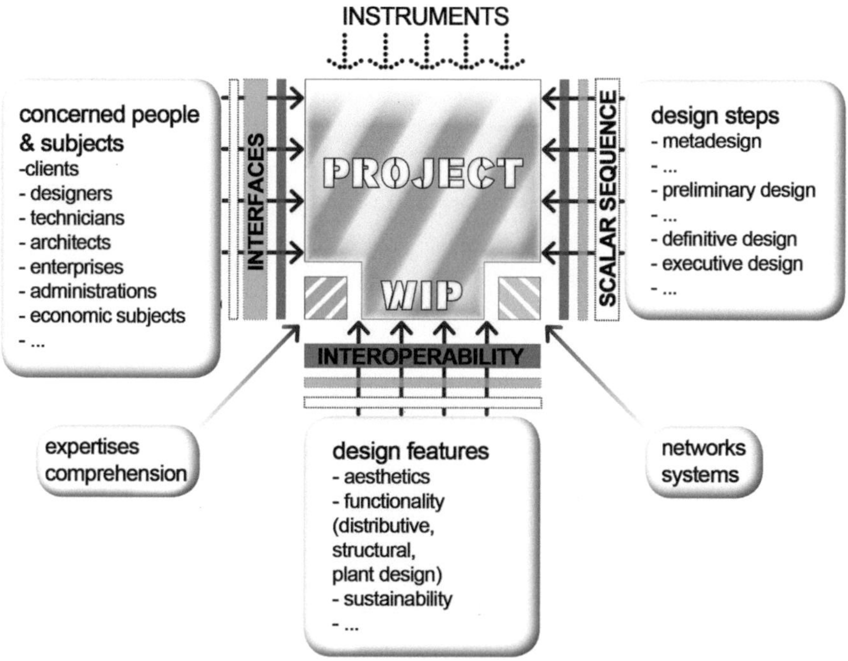

Fig. 5.7 Work in progress design process and digital concepts—elaborated with Casetta (2009)

In an integrated approach to design and construction, the process allows the connection of different technical figures and models, according to a vision capable of using and interconnecting innovative software solutions such as parametric processes, algorithm-based optimisation, BIMS and digitisation of production processes. Recent connections between dynamic energy analysis software, such as *EnergyPlus*, and analysis and design environments allow us to glimpse new design scenarios integrated with energy needs and sustainable performance, particularly in view of a higher scale vision than the single integration of a technology in a building. Figure 5.9 describes the integration between functional analyses and environmental requirements through a scheme developed in 2008 during the master dissertation. This scheme can be furthered with new approaches capable of integrating the surrounding context in order to achieve neighbourhood sustainability, which is necessary to design true urban resilience and an approach aimed at net positive design. Functional requirements, which represent the results of the analyses, are directly related to environmental requirements, in compliance with an approach linked to the spatial and climatic constraints of the site. The assessment methods are integrated directly into the design, translating the systems to weights and indicators in an explicit algorithmic vision capable of replacing the linear process in a dynamic and interactive approach. It is therefore possible to integrate environmental needs and specifications from the earliest stages of design development into a dynamic and integrated real-time performance visualisation, not based on an ex-post design assessment or support

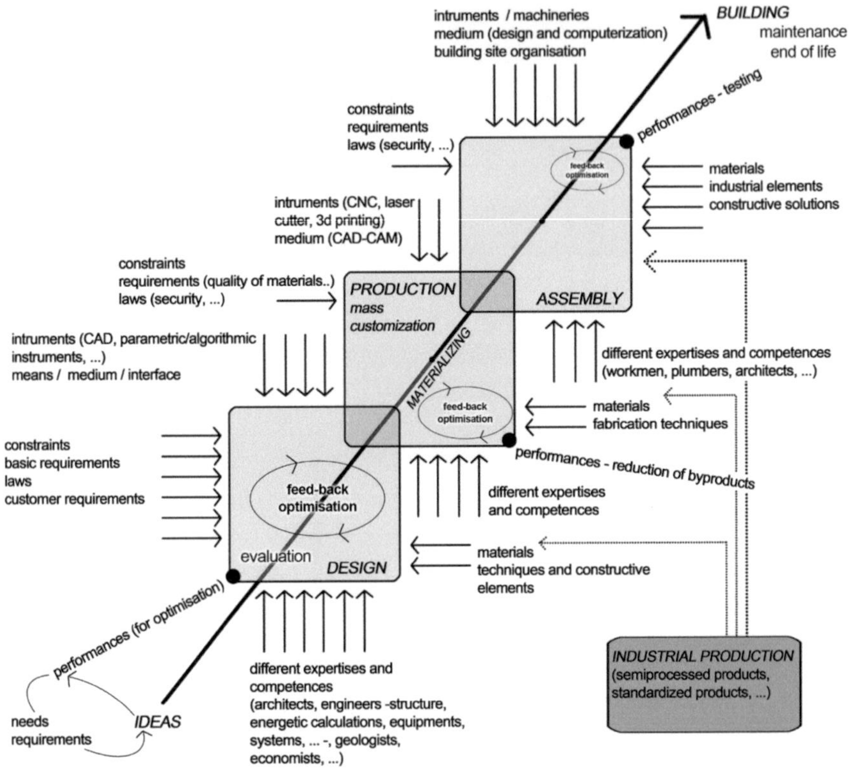

Fig. 5.8 Schematic representation of the innovative digital approach to the architectural design process. Materials, instruments and tools, constraints, and technical knowledge connected to the project—re-elaboration from Chiesa (2013)

or one which is independent of the main geometric information flow. This approach also allows the algorithmic management of the generation of shapes, maximising the response according to specific parametric requests, using a genetic algorithm in Galapagos, for example, and integrating it with Rhinoceros for the geometry and specific programs for analysis, including EnergyPlus, DIVA, Ecotect and FEM analysis, as outlined in previous steps. The transition from an assessment to a geometry can be easily implemented in a parametric environment, also forcing the explanation of choices and promoting conscious planning, because constraints and needs are transformed into generative parameters. The analysis/assessment tools must also be transformed into parametric models. The site's microclimatic matrix tool (e.g. Brown and Dekay 2001; Grosso 2008), for example, capable of assessing the performance anticipated in the various areas of the site in relation to exposure to/protection against the sun and wind, can be automated. The automatic calculation of shadowing dynamics is easy to achieve in the current design platforms thanks to internal plug-ins or, where possible, by developing a script based on the calculation formulas of the apparent solar azimuth and height angles in the place under examination on

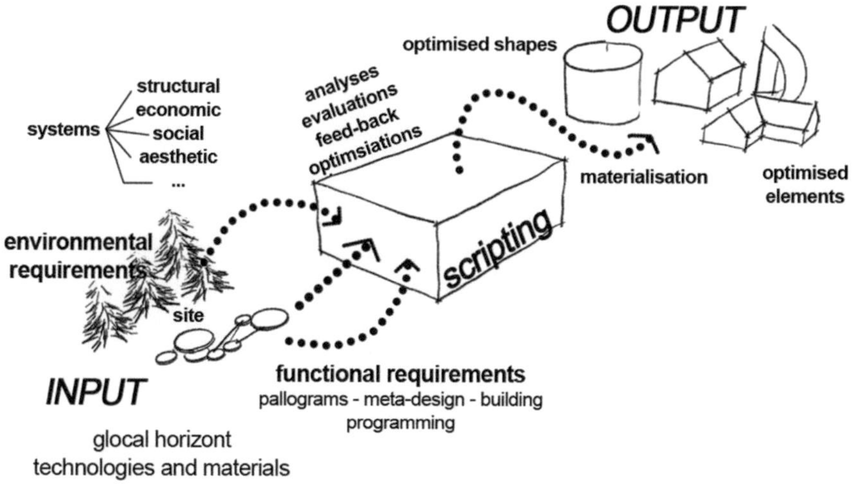

Fig. 5.9 Concept to integrate sustainable, environmental, functional and technological aspects in the parametric design process (Chiesa 2013)

different days of the year and at different times of day (e.g. Grosso 1986). Diversely, the calculation of the size of the wind wake core zones—in which the reduction in wind speed due to obstacles is higher than 50% of the initial value—can be carried out using CFD (Computational Fluid Dynamic) simulations or fitting curves derived from databases previously generated in CFD environments and wind tunnel testing (e.g. Chiesa and Grosso 2015)—see the script for Grasshopper STEMMA tool (e.g. Chiesa and Grosso 2019).

Also on the subject of geometric-generative systems, it is interesting to mention the *9-Tsubo_House* competition in 2006 in which Matsukawa and 000studio (2006) presented a design for an algorithmic space where the building, installed in a 30 m^2 lot flanked by a twin lotus used to house a wood, took shape independently, following the writing of a software capable of analysing the site, the context variables and the functional needs in a game. The software was able to generate infinite evolutionary structures that responded to the need-driven framework and performance requirements. The building had to be built using timber obtained from the lot, in an alternation between the lot occupied by the building and the lot planted with trees dictated by the moment in which the trees were ready to be felled, simulated and calculated by computer in consideration of the shape of the building and the amounts of timber needed to obtain the pieces for its construction, including waste and scrap. This idea, although outdated in the field of digital design, draws attention to the potential of new tools and management platforms, which become the real design space in the world of digital ages. The choice, in a Darwinian perspective, of the best solutions between datasets generated on the basis of specific algorithms, brings the natural and artificial worlds back to the axis of hybridization (Molinari 2006).

5.4 Conclusions

The platform can be set up as a place of interaction between the different technical, computerised, instrumental and need-driven components present in the modelling and materialisation processes of complex design. The building as a materialisation of digital information derives from a precise quantification of the design and construction content of a project, defining complexity as the ratio between the amount of additional design to the amount of additional construction (Mitchell 2005). The architecture of the digital ages allows high levels of complexity which were previously unknown, enabling outputs that are more responsive to specific needs, to the local reality and to design intentions, in a more sensitive perspective than was possible in the modernist and post-modern ages. It is, however, essential to take into account the way platforms are built and operated in order to limit any problems that may arise. First of all, as already pointed out during the chapter, it is essential to choose the right degree of openness of the platform to allow intervention on definitions and explanations, implementing process and design space, where necessary. It is also important to verify and perfect interoperability between different software products, tools, languages, models and platforms. Poor interoperability of digital design and construction platforms poses a serious limitation to the applicability of the tools available and digitally mediated design. The proprietary nature of platforms is another key issue in view of the fact that they are influenced by the action and operation of data and information as well as materialisation and modelling processes that could send bits to third parties. This proprietary nature may also affect interoperability and policies for developing or implementing the potential of the design space and, consequently, the design method itself. Lastly, aspects related to the incidence of updates and changes in ownership or management of the developers of plug-ins or sub-platforms, which may implicate the updating of the entire chain of software platforms and sub-platforms and, at the same time, change the way in which scripts are organised, are essential. There is also a risk that scripts that have already been pre-compiled will become obsolete. Even though an open source version is available, the acquisition of Pachube by Xively has interrupted the ease of use of the platform in terms of interoperability with other design areas, like some specific Grasshopper applications related to Google-based or Arduino-based implementation systems. This has significant repercussions, especially in cases where the online platforms for sharing and collecting data are connected to nodes developed by individual users that cannot be quickly updated.

In view of the new programmes and challenges of sustainability and energy saving, with a view to 2020 but, above all, to net positive building and district, it will become increasingly necessary to overcome the traditional implicit process between knowledge, tools and design. Despite current awareness that the sum of sustainable technologies does not translate into a sustainable home and that the juxtaposition of sustainable buildings does not create a sustainable neighbourhood, city or region, it is necessary to develop integrated tools capable of directing specific local projects towards shared net positive horizons. The choice of the scale of sharing must be able to vary from parameter to parameter and allow design which is increasingly linked to algorithms and growing complexity, which can no longer be managed implicitly, if not at the expense of quality and mismatches between the operating and the asset rating. Monitoring and data production techniques, linked increasingly to correlations and statistical analyses, will make it possible to obtain evaluations and knowledge that increasingly overlap with reality. Nevertheless, it is still necessary to know how to grasp them and integrate them into design, whether preliminary, definitive, executive, on site, maintenance-related or aimed at management. This concept embraces a vision that places the work in progress of the model and reality at the centre of a vision based on interoperability and interfacing, in a holistic approach between different experiences and knowledge, networking, actions on different scales and diversified and specialised tools.

References

Aliberti A, Bottaccioli L, Cirrincione G, Macii E, Acquaviva A, Patti E (2018) Forecasting short-term solar radiation for photovoltaic energy predictions. In: Proceedings of the 7th conference on smart cities and green ICT systems (SMARTGREENS 2018), Madeira, Portugal, 16–18 Mar, pp 44–53

Anderson C (2012) Makers: the new industrial revolution. Random House Business Books, London

Bochicchio D et al (2010) C# 4. Guida completa per lo sviluppo. Hoepli, Milano

Brown GZ, Dekay M (2001) Sun, wind & light. Wiley, New York

Casetta E (2009) Parametri e architettura temporanea: progettare il prototipo. Master Degree thesis, MArch in Architecture, Politecnico di Torino, Academic Year 2010/11, supervisor: Pagani R, co-supervisor: Chiesa G

Chiesa G (2010) Biomimetica, tecnologia e innovazione per l'architettura. Celid, Torino

Chiesa G (2013) M.E.T.R.O. (Monitoring energy and technological real time data for optimization) innovative responsive conception for city futures. PhD thesis, Politecnico di Torino, Torino

Chiesa G (2015) Paradigmi ed ere digitali. Il dato come parametro di innovazione in architettura e urbanistica. Accademia University Press, Torino

Chiesa G (2017) Biomimetics. Technology and innovation for architecture. Celid, Torino

Chiesa G (2019a) Environmental design strategies in different density-urban contexts. TECHNE 17:183–190

Chiesa G (2019b) Optimisation of envelope insulation levels and resilience to climate changes. In: De Joanna P, Passaro A (eds) Sustainable technologies for the enhancement of the natural landscape and of the built environment. Luciano Editore, Napoli, pp 305–338

Chiesa G, Grosso M (2015) Accessibilità e qualità ambientale del paesaggio urbano. La matrice microclimatica di sito come strumento di progetto. Ri-Vista 13(1):78–91

Chiesa G, Grosso M (2019) A parametric tool for assessing optimal location of buildings according to environmental criteria. In: Sayigh A (ed) Sustainable building for a cleaner environment. Springer, Cham, pp 115–130

Chiesa G, Huberman N, Pearlmutter D, Grosso M (2017a) Summer discomfort reduction by direct evaporative cooling in Southern Mediterranean areas. Energy Procedia 111:588–598

Chiesa G, Simonetti M, Ballada G (2017b) Potential of attached sunspaces in winter season comparing different technological choices in Central and Southern Europe. Energy Build 138:377–395

Chiesa G, Grosso M, Acquaviva A, Makhlouf B, Tumiatti A (2018) Insulation, building mass and airflows—provisional and multi-variable analysis. SMC—Sustainable Mediterranean Construction 8:36–40

Chiesa G, Acquaviva A, Grosso M, Bottaccioli L, Floridia M, Pristeri E, Sanna EM (2019a) Parametric optimization of window-to-wall ratio for passive buildings adopting a scripting methodology to dynamic-energy simulation. Sustainability 11:3078. https://doi.org/10.3390/su11113078
Chiesa G, Huberman N, Pearlmutter D (2019b) Geo-climatic potential of direct evaporative cooling in the Mediterranean region: a comparison of key performance indicators. Build Environ 151:318–337
Devoto G, Oli GC (2009) Devoto-Oli Il vocabolario della lingua Italiana, edizione 2009. Le Monnier, Firenze
Grosso M (1986) Dinamica delle ombre. Celid, Torino
Grosso M (2008) Il raffrescamento passivo degli edifici in zone a clima temperato, 2nd edn. Maggioli, Sant'Arcangelo di Romagna
Grosso G, Acquaviva A, Chiesa G, da Fonseca H, Bibak Sareshkek SS, Padilla MJ (2019) Ventilative cooling effectiveness in office buildings: a parametrical simulation. In Proceedings of the 39th AIVC—7th TightVent & 5th venticool conference—smart ventilation for buildings, Antibes Juan-Les-Pins Conference Centre, France, 18–19 Sept 2018, pp 780–788. ISBN 2-930471-53-2. Available at https://www.aivc.org/download/aivc2018-proceedings.pdf, last view May 2019
Huberman H, Pearlmutter D, Gal E, Meir IA (2015) Optimizing structural roof form for life-cycle energy efficiency. Energy Build 104:336–349
Košir M, Gostiša T, Kristl Z (2018) Influence of architectural building envelope characteristics on energy performance in Central European climatic conditions. J Build Eng 15:278–288
Matsukawa S, 000STUDIO (2006) Algorithmic space (9-Tsubo_House). In: Hwang I et al (eds) Verb natures, architecture Boogazine. Actar, Barcelona
Mayer-Schönberger V, Cukier K (2013) Big data. Una rivoluzione che trasformerà il nostro modo di vivere e già minaccia la nostra libertà. Garzanti, Milano [or (ed) Big data: a revolution that will transform how we live, work, and think. Houghton Mifflin Harcourt, Boston]
Mitchell WJ (2005) Construction complexity. In: Martens B, Brown A (eds) Computer aided architectural design futures 2005. Springer, Netherlands, pp 41–50
Molinari E (2006) In: Onden L (ed) Invenzioni bioispirate. Explora la Tv delle Scienze, Rai Educational, 28 Apr 2006
Nakano A (2015) Urban weather generator user interface development: towards a usable tool for integrating urban heat island effect within urban design process. M.S. thesis, MIT Building Technology
Nakano A, Bueno B, Norford LK, Reinhart C (2015) Urban weather generator—a novel workflow for integrating urban heat island effect within urban design process. Build Simul 2015. http://urbanmicroclimate.scripts.mit.edu/publications.php, last view Dec 2018
Osello A et al (2013) Multidisciplinary team activity using BIM and interoperability. A Ph.D. course experience at Politecnico di Torino. In: Gambardella C (ed) Heritage architecture and design. XI International Forum Le vie dei Mercanti, Aversa-Capri, 13–15 June 2013. La Scuola di Pitagora editrice, Napoli, pp 880–889
Osello A, Acquaviva A, Del Giudice M, Patti E, Rapetti N (2016) District information models. The DIMMER project: BIM tools for the urban scale. In: Pagani R, Chiesa G (eds) Urban data. Tools and methods towards the algorithmic city. FrancoAngeli, Milano, pp 231–261
Pagani R (ed) (2009) BdS 2040—challenge all energy. TAO—Transm Archit Organ 1:24
Palme M, Inostroza L, Villacreses G, Lonato A, Carrasco C (2017) From urban climate to energy consumption. Enhancing building performance simulation by including the urban heat island effect. Energy Build 145:107–120
Perini K, Chokhachian A, Dong S, Auer T (2017) Modeling and simulating urban outdoor comfort: coupling ENVI-Met and TRNSYS by grasshopper. Energy Build 152:373–384
Tibbits S et al. (2011) Python101 for Rhinoceros 5. http://www.rhino3d.com/download/IronPython/5.0/RhinoPython101, last view 2013

Weinberger D (2012) La stanza intelligente. La conoscenza come proprietà della rete. Codice edizioni, Torino [or (ed) (2011) Too big to know: rethinking knowledge now that the facts aren't the facts, experts are everywhere, and the smartness person in the room is the room. Basic Book, New York]
http://energyplus.net/, last view Dec 2018

Chapter 6
Data, Properties, Smart City

Big data mark the moment when the "information society" finally fulfils the promise inherent in its name. Data come to occupy centre stage.
Mayer-Schönberger and Cukier (2013, 258)

This chapter introduces the concept of the third digital age, a development of the first and second ages, and focuses on the centrality of data in design and in urban processes. The topic of the platform is used to analyse the repercussions and risks typical of design spaces based on the centrality of data, especially from the point of view of those who possess and interact with big data. When IT is configured as a double parameter of influence on the city and on the design dynamics, ownership of data becomes a nodal point for the management of urban spaces and processes, whether they concern modelling or materialisation. Data collection methods play an important role in defining the enabling and inclusive value of ICT or, on the contrary, the risk of disabling, non-inclusive or limiting drifts. Lastly, it analyses the importance of the choice and organisation of datasets and algorithms capable of assigning proprietary organisations and related technologies the role of tool and not of decision-maker to big data, introducing the concept of human use of human beings (Wiener 1966).

6.1 Third Digital Age

Buildings were once materialized drawings, but now, increasingly, they are materialized digital information

Mitchell (2005, 41)

When the floor is datised there is no ceiling to its possible uses

Mayer-Schönberger and Cukier (2013, 131)

The concept of the first digital age, defined by Oxman (2006), can be used as a key to reading and theoretically programming digitally mediated design focused on

G. Chiesa, *Technological Paradigms and Digital Eras*,
PoliTO Springer Series, https://doi.org/10.1007/978-3-030-26199-3_6

the modelling process and software innovations. The second digital age, on the other hand, is defined as a key to reading and theoretical programming digitally mediated design for construction and the construction process itself, linked to materialisation and hardware innovations. The third digital age focuses on theoretical programming and innovation of digitally mediated design, construction, management, analysis and optimisation methods, derived from an approach based on information and complexity. This innovation concerns both hardware and software and is linked to the Internet of Things (IoT).

6.1.1 Datisation

A fundamental point of innovation in the relationship between ICT and society lies in the concept of datisation understood as the ability to extract data from materials that, until just recently, were considered totally worthless or were not convertible into quantitative forms. Datisation includes the production and subsequent analysis of data to convert them into information. Datisation is a concept that already existed "many centuries before the digital age [and is] quite different from digitisation, the process by which digital information is converted into the numbers zero and one of the binary code" (Mayer-Schönberger and Cukier 2013). It is clear, however, that the digital revolution plays a fundamental role in innovation related to big data. "Today's information systems definitely make big data, which are actually part of man's constant attempt to measure, understand and analyse the world, possible" (Mayer-Schönberger and Cukier 2013, 110). This statement is directly related to the concept of world quantification, a process that is progressively invading our way of life, which makes it increasingly possible to measure elements and concepts that seemed unmeasurable until not so long ago. An increasing number of aspects of the world around us and of our way of life are being measured or can be converted into data, and this means that the "next frontiers of data processing are more personal: they are related to our relationships, our experiences and our moods. The concept of datisation is the backbone of the social media that thrive on the web" (Mayer-Schönberger and Cukier 2013, 127).

For example, it is possible to use geo-localisation to spatially analyse data extracted from information posted on social networks and other online communication channels such as Facebook, Twitter and Flickr, through special plug-ins compatible with geometric programs and databases. The processes of datising information found on social networks and other sites are not based solely on the content of posts and attachments, but also on the metadata linked to them (Figs. 6.1, 6.2, 6.3, 6.4, 6.5, 6.6, 6.7, 6.8 and 6.9). A tweet is a particularly precious element because it has a very rich and articulated meta-information based on 33 separate elements, including the geographical position. The use of programs capable of managing these datapoints, making their respective datasets available for analysis, is going to dramatically change the way we manage different phenomena. It is possible to read information related to social networking posts using applications capable of han-

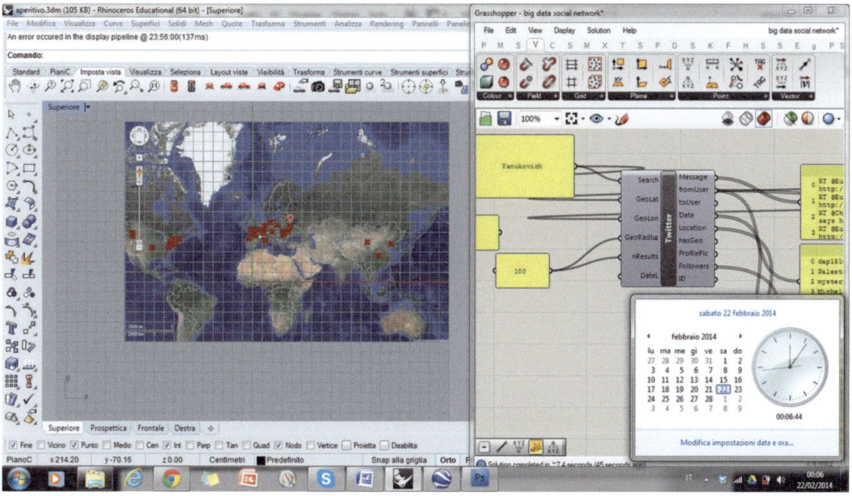

Fig. 6.1 Geopositioning analysis based on tweets reporting the "Yanukovych" keyword—00:06 a.m. on the 22nd of February 2014

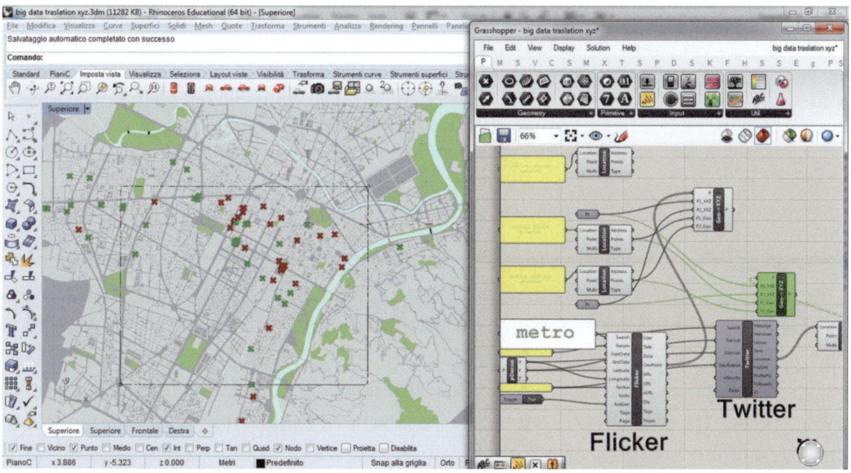

Fig. 6.2 Maps of geolocalised tweets (green) and Flickr images (red) referring to the tag "metro" for the city of Turin (30 km radius around Piazza Castello—one of the central squares)

dling bee codes, including *Mosquito*, a Grasshopper plug-in developed by Carson Smuts. Thanks to these applications, it is possible to geolocate twitter information, identify specific IDs, analyse the number of followers of certain IDs or specific posts in order to build maps and interpretations aimed, for example, at locating urban phenomena or geographically managing the construction of mental maps. It is also possible to spatially analyse the photographs posted on Flickr, automatically archive

Fig. 6.3 Maps of geolocalised tweets (green) and Flickr images (red) referring to the tag "Mole" for the city of Turin (30 km or ray around Piazza Castello—one of the central squares). The Mole is the symbol of the city of Turin. See how red points are localised near the monument and connected to pictures, while the green ones are spread around the city level and are connected with written messages

them and use this information for the development of new territorial analyses linked to the use of specific services, flows and perceptual phenomena—see also Sect. 7.1.2.

This has profound repercussions on the possibilities offered by digital media in innovating the tools for decision making and analysis of phenomena, particularly for design purposes. The research of the Senseable City lab, especially that reported in Ref. (Girardin et al. 2007) constitutes a database of possible applications in this sense. Lynchian analyses of inhabited space, enriched by new ways of developing mental maps and perceptual and visual analyses of space, are accompanied by new potentialities for the study of landscape, which work on perception and movements (Chiesa and La Riccia 2013; Chemollo and Orsenigo 2004). The use of specialised photography, conceived as a design tool for delegation to specific professional figures, hybridising different and essential types of knowledge, finds a key to innovation, especially for the implications in territorial and landscape analysis, in these digitally mediated tools (Chiesa and Di Gioia 2011; Chiesa and La Riccia 2011). The perceptive readings of the naive realism of the professional photographic eye take place mainly with the datisation of sentimental geographies (Ghirri 1997) and memory (Augé 2004), capable of restoring dignity and identity to living spaces, like a "sort of recomposition of family albums" (Lupano 1989).

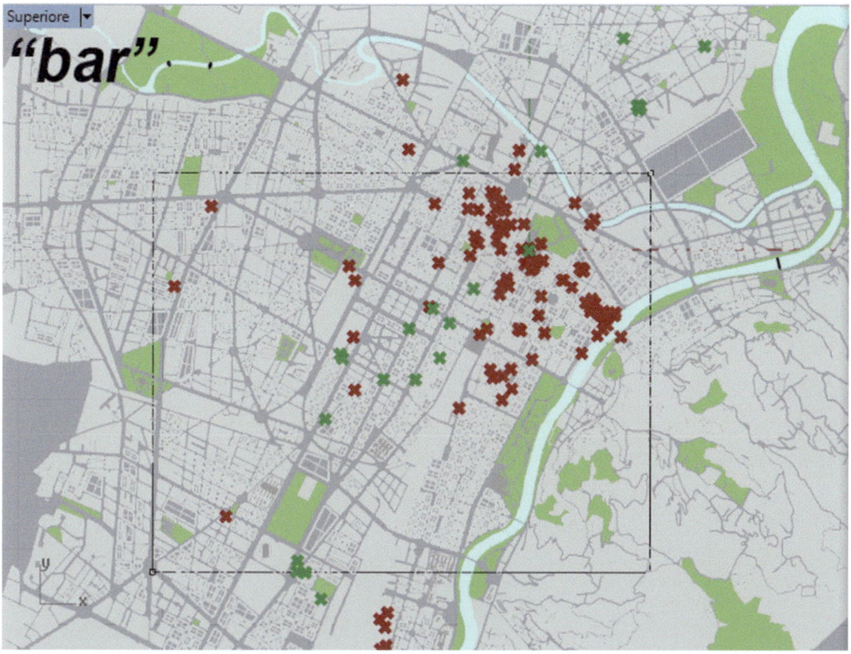

Fig. 6.4 Maps of geolocalised tweets (green) and Flickr images (red) referring to the tag "bar" for the city of Turin (30 km radius around Piazza Castello—one of the central squares)

6.1.2 Potential Applications of Social Network Datisation: Spatial Implications (Facts; Keywords; Landscape Perception Value Maps)

In this paragraph, sample applications based on the usage of datisated information from social networks connected with CAD/GIS environments are reported. These examples are only a small introduction to the large potential that these implications may have on urban design and strategies definitions.

Firstly, the use of the grasshopper plug-in Mosquito is applied to the study of socio-political phenomena, in order to define spatial images of an event. In the example reported in Fig. 6.1 a spatial analysis of data derived from the datisation of tweets' metadata (internet 2.0) is reported.The image shows the geopositioning of different tweets containing the word "Yanukovych" during the day of his removal as the prime minister of the Ukraine, time 00:06 a.m. on the 22nd of February 2014. The map is based on the processing of 500 tweets (100 for each parallel process) by analysing their geographic metadata at world level with a buffer around the 50° North parallel. The map was imported to the Rhinoceros environment by using a Google re-elaboration engine. On the other hand, Figs. 6.2 and 6.3 shows the geopositioning of tweets around the Майдан Незалежності square using respectively "Yanukovic" and "Tyniachenko" as keywords.

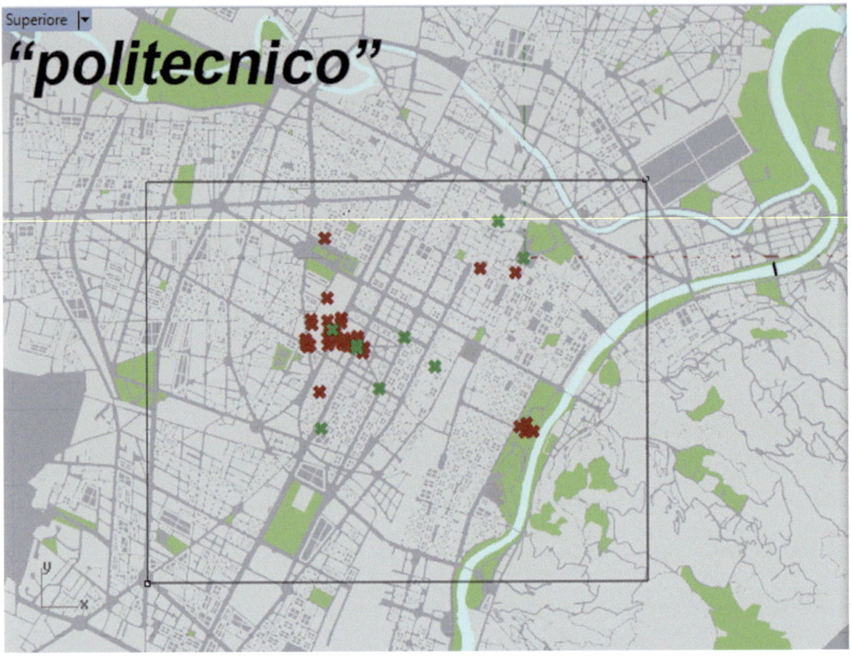

Fig. 6.5 Maps of geolocalised tweets (green) and Flickr images (red) referring to the tag "politecnico" for the city of Turin (30 km radius around Piazza Castello—one of the central squares). Politecnico refers to the Politecnico di Torino (University). Red points identify two of the major location sites of this university in the city (Corso Duca with campus and Valentino Castle)

Secondly, the same approach to datisation is applied to the geo-definition of urban keywords. In particular, in Figs. 6.2, 6.3, 6.4, 6.5, 6.6, 6.7 and 6.8, the localization on the map of the Municipality of Turin of data derived from datisation of information obtained from Internet 2.0 sources (Twitter and Flicker) are reported. Points have been translated by the WGS84 coordinate system (used by several geographic sources, including Google) into an XYZ system which is adaptable to geometries and maps in the CAD environment in order to be able to view them in the correct position. This step should be performed differently depending on the software used, for example ArcMap and QGIS may use the WGS82 system directly.

These figures are based on a map which is imported from the SIT (Territorial Information System) of the Municipality of Turin into the Rhinoceros environment. Thanks to the Mosquito plug-in connected with Grasshopper some data and metadata of tweets and images from Flicker have been collected in databases with respect to specific Flicker keywords—tags. Figures show the results derived from Flicker in red and the Twitter data in green. The search keyword is shown in each image. As underlined by figures, Twitter data are geographically less precise but theoretically more suitable to populate a map of personal sensations. The Flicker data, on the other hand, can help identify a specific location and functions with high geographical precision. The same source can also be used in geo-locating points of interest, for

Fig. 6.6 Maps of geolocalised tweets (green) and Flickr images (red) referring to the tag "parco valentino" for the city of Turin (30 km radius around Piazza Castello—one of the central squares). The specific area of the park (Valentino) is perfectly identified by the distribution of Flickr points

example of a tourist nature, and areas of incidence/perception of particular buildings or functions. In fact, a photograph tends to be located at the capturing point whereas a tweet is geo-referenced at the point where it is written, although many users do not provide a specific address but only the name of the city. Furthermore, the analyses conducted on Twitter refer to almost real time readings while those on Flicker are based on longer periods and barely allow us to identify short-term temporal variations. Twitter data can be implemented with analyses based on the amount of followers of the user to whom the datapoint is associated, in a potential scoring perspective of the data.The last image (Fig. 6.8) refers to the Twitter map, scored on the number of followers, of the word "Torino" tweeted in a radius of 30 km from Piazza Castello—one of the central squares in the city. Radiuses of the presented circles are, in fact, proportional to the number of subscribers to the channel of who tweeted the keyword. It is interesting to note how this analysis may help to identify both specific points and cultural usage in a city to get tourist information early on meeting and leisure areas in a new city.

Finally, the use of datisation approaches may be applied to the definition of landscape perception value maps in order to implement, for example, the definition of landscape preservation areas connected to monuments and visual perceptive value preservations. This approach is directly connected with landscape constraints in

Fig. 6.7 Maps of geolocalised tweets (green) and Flickr images (red) referring to the tag "aperitivo" for the city of Turin (30 km radius kparound Piazza Castello—one of the central squares). The word "aperitivo" denotes a specific way of having dinner (finger food) in the central parts of the city which are associated with the "movida" (lively night life). Red points identify two of the main areas where aperitivo bars are located (Piazza Vittorio and Quadrilatero latino)

tourist areas, e.g. by limiting the installation of PV panels or other RES (renewable energy systems) in the proximity of monuments, landscape or preservation areas. Although it is important to preserve landscape value perception and visibility, radius approaches may protect surfaces (e.g. roofs) that are not directly visible by users (people), and/or not preserve environmental spaces which are distant from the protected elements, but very visible in its environment. Potential approaches to overcome this problem by protecting only effective landscape portions and allowing the space of application of RES to enlarge, can be constituted by 3D-gis analyses such as the viewshed instrument—see for example Chiesa and La Riccia (2016) and La Riccia (2017). Nevertheless, the definition of analysis points to apply mutual visibility analyses (e.g. viewshed) is still an open question—Historical panorama? Nearest open spaces? Inverse analysis by the centre with a fixed diameter? The possibility to use datisation from social networks by identifying through geo-points where users enjoy the specific landscape valuable element is a valid starting point to propose alternative solutions. An early study based on the photograph database collected in Flickr by users was reported by the author (Chiesa and La Riccia 2016). Furthermore, it is possible to suggest that this may also be conducted by implementing a filter mask in other social networks such as Facebook and Twitter, in which the iconographic

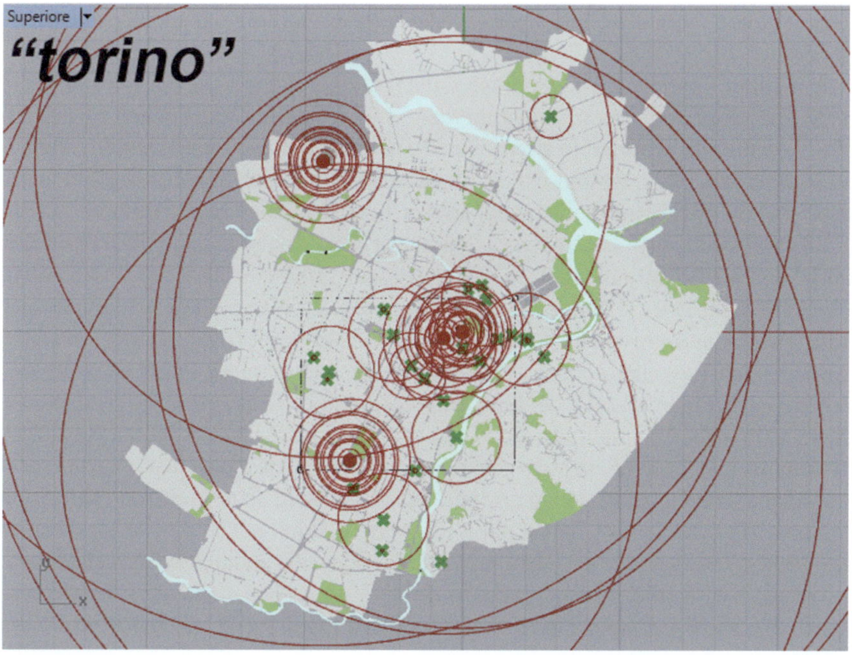

Fig. 6.8 Maps of geolocalised tweets (green) and Flickr images (red) referring to the tag "torino" (Turin) for the city of Turin (30 km or radius around Piazza Castello—one of the central squares). The main circles are around the city centre (around Piazza Castello) and the two stadiums

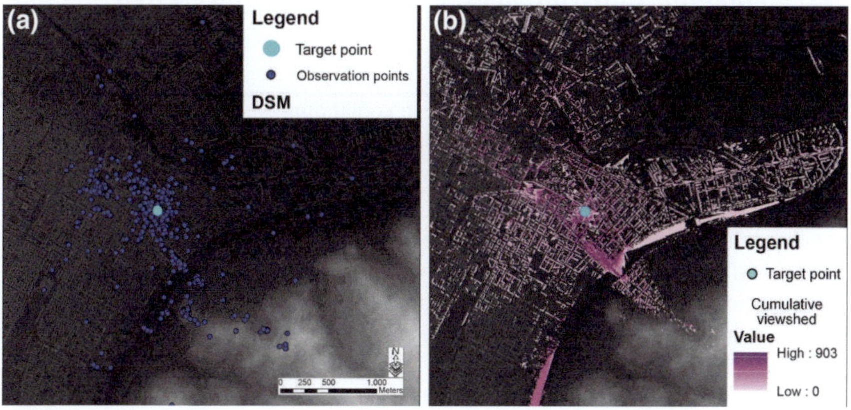

Fig. 6.9 Sample application of the cumulative viewshed approach to landscape perceived value mapping by using datisation from social network (Flickr). **a** Map of observation points; **b** cumulative viewshed results

context is not so evident, but other aspects such as cultural and social dimensions are.

A method to apply the iconographic database georeferred in Flickr to map the landscape value perception and further use it to define the local priority to install RES systems was reported in a recent study (Chiesa 2017b). This example refers to the overlapping of perception indicators based on the calculation of weighted raster maps while considering both points spatial intensity maps (e.g. Heatmap tool in QGIS) and multi-visibility analyses. Points for these analyses can be derived from tourist and historically recognized sites or directly from social network datisation. When the point layer is defined, it can be used to perform not only spatial intensity indicators (which are useful to define multi-landscape defence priority) but also viewshed analyses that connect each point with one or more landscape valuable elements (e.g. landmarks) by defining the point direction to target(s), the distance, the connected viewshed amplitude according to photograph lens translation. Results of all analyses may be overlapped to generate a cumulative viewshed that may be finally overlapped to PV potential map e.g. performed by using solar radiation calculation tool on a 3D layer in GIS. These approach may be summarised in 4 steps: (i) point definition; (ii) target and viewshed inputs definition; (iii) cumulative viewshed map; (iv) overlap with RES potential map to define a priority map for RES installation. See an extract of a sample application reported in Fig. 6.9 and adapted from Chiesa (2017b).

6.1.3 Potential Consequences: Data, Models, and Design Issues

The possible repercussions of these tools on social analysis, accessibility and planning are many, especially in cases in which they are aimed at the design of urban spaces, interfacing different components which it was very hard to measure until just recently. Think of real-time travel, modes of transport, commercial choices and the moods related to certain keywords. This is new information capable of interfacing with other different data in compliance with increasingly complex patterns of analysis. It is possible, for example, to link datasets from *data exhaust* analyses with parametric analysis of shapes and uses of urban spaces and services (Sevtsuk and Amindarbari 2012).

However, a new frontier of retrieval and use of data and information is just around the corner. If the web 2.0 has allowed the connection of people and the datisation of a huge amount of new information which was previously intangible, the web 3.0 will allow the transformation of any object into a sensor and a latent actuator. Along with the datisation of everything around us, "the value of data [will shift] from primary use to possible future uses" (Mayer-Schönberger and Cukier 2013, 137). The exponential increase in the amount of data that can be stored [Weinberger (2012); even though, as we are reminded by Floridi (2013)—so much information will be

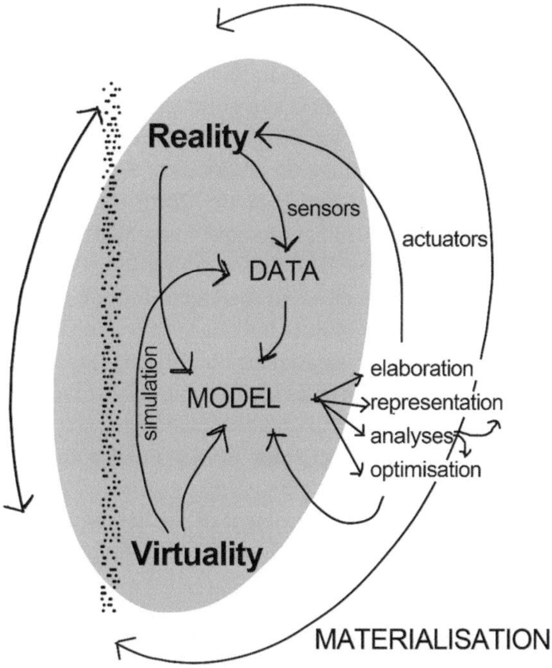

Fig. 6.10 Schematic representation of potential implications of the datisation process

produced that it will be impossible to store it all] will lead to an increase in the amount of information stored, which will result, on the one hand, in a radical change in the concept of memory, the limit of which is no longer met by the amount of data stored (Mayer-Schönberger and Cukier 2013), and, on the other hand, the possibility to use information several times, increasing its marginal value (Weinberger 2012). The amount of information produced will be so large that, as theorised by Floridi, it will not be possible to store it all. Indeed, in 2013 alone, the estimated amount of data stored was 1200 Exabytes, of which only 2% in paper format. This reasoning has important implications on how we collect and define the types of data to be monitored and introduces the concept of extensibility: "it is advisable to collect as much data as possible, and make it extensible, taking into account future secondary uses from the start" (Mayer-Schönberger and Cukier 2013, 150). The reuse of data and the problem of their progressive obsolescence also concerns the field of architecture, where the designer can reuse part of the pre-coded content in the design process, possibly enriching it with new inputs from meta-design analysis and real-time data (Mitchell 2005). At the same time, the need to archive as built digital models of buildings is becoming increasingly evident, including as much information as possible that can be connected to them from a BIM point of view, including information relating, for example, to materials, maintenance scheduling and consumption (expected, monitored subsequently, …).

In an article dated 2005, William Mitchell, a professor at mit, says that buildings are becoming materialised digital information. Complexity, linked to the relationship between the increase in design content and the corresponding construction, production and assembly content, is the basis of what Oxman defined, in 2006, as the future digital age. In the space of digitally mediated design, the modelling and materialisation processes of the third age will tend to position and intercept information flows, drawing parameters for the generation, optimisation, evaluation, visualisation and analysis of performance. In addition, as Tim O'Reilly points out, data is a platform in which datapoints and datasets "form the basis for new products and business models" (Mayer-Schönberger and Cukier 2013, 166). The extensibility of data and information is a key element for digitally mediated design, because the same data, read on different scales and mediated through the construction of specific datasets on time and geographical intervals, can enter the process of design, construction and maintenance as a parameter of different algorithms, whose purposes may differ significantly. On this basis it will be possible to make optimisations and assessments on specific architectural choices, on design on a district scale, on analysis aimed at urban policies and decisions, from a much more aware and advanced perspective (Figs. 6.10 and 6.11).

The model, as a tool, subject of the design action and focused representation of reality will also be able to interact with the real world thanks to IoT platforms and

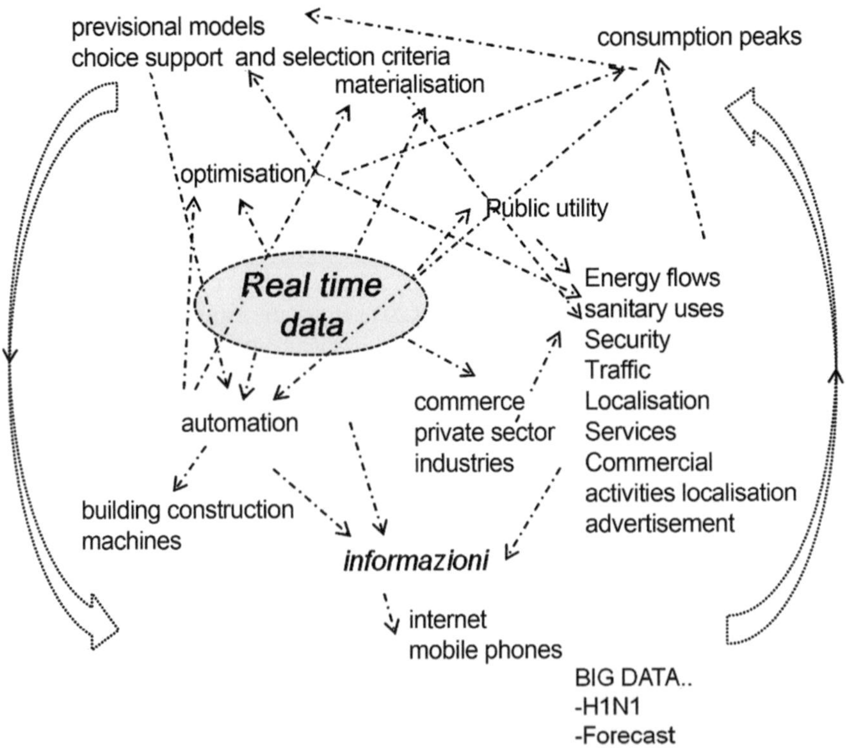

Fig. 6.11 Schematic representation of some implications of real-time data production

real-virtual interaction tools linked to sensors and actuators, increasing its affinity with reality and its intelligence with a view to creating new real-virtual interconnections. Models, materialisations and reification, whether real or virtual, become nodes of platforms and networks of information and data. Real objects tend to become physical models of more complex realities, where even "things" become intelligent and embedded, in a context where strokes are enveloping and integrated. Real-time data changes the implications of the future digital age even more, transforming sensors, monitoring, modelling, analysis, actuator systems and materialisation processes into dynamic phenomena. The cities of the future can be planned parametrically on the basis of multi-temporal urban analyses, capable of focusing on different aspects, even after the data has been acquired, as happens in recent photography techniques, where the choice of focus takes place during post-production. Market analyses, flows and the use of services can be radically modified, as hypothesised by the research coordinated by Sevtsuk at the City Form Lab of the Polytechnic of Singapore, on processes that, in the vision defined here as third age, take place on different scales of design.

6.2 Cities and Data—Decisions' Schemes and Urban Properties

> Whether we entrust our decisions to metal machines or living machines that are offices, large laboratories, armies or industrial companies, we will never get the right answers to our questions unless we ask the right questions.
>
> Norbert Wiener 1950 [1966]

The construction of models, their population, parametric analysis, information management and implementation response, which can be traced back to the concept of modelling, management of the modelling platform and subsequent materialisation, involve administrative, economic, scientific and ethical decisions. These connections become more complex and articulate in the world the third digital age, when IT has a direct influence on design decision-making platforms, with their enabling power, etc. The way the digital age is managed, affecting information, technologies and communication methods, opens up a number of questions and risks associated with the use of data and technologies themselves.

In particular, studies carried out from the birth of digitisation up to the 1970s should now be rediscovered and reinterpreted in the light of the changes that are taking place in the way scientific research is understood and the change not of paradigm, but of paradigms (Mayer-Schönberger and Cukier 2013; Nielsen 2012; Kuhn 1969).

It is interesting to further the topic of the risk of human use of human beings, which reminds us how a "human individual represents an expensive investment in study and culture [but it is possible to develop particular forms of organisation, such as to induce] the disposal of the enormous advantage of education and use human material" to perform limited functions, as happens with ants (Wiener 1966). The progressive dissemination of an infinite memory, lacking in the phase of elaboration intrinsic in memorisation and oblivion, a concept studied by Tosoni (1992), and the progressive dissemination of models capable of representing elements that could not be given before, risk increasing the machinist drifts of the use of human beings. Although big data allow developments that may escape the ability of the individual to influence technology, it is essential to emphasise the importance of the mature and conscious human–machine relationship, which stems from "knowing what we want; and, therefore, determining not only how to achieve our goals, but also what our goals should be" (Wiener 1966, 225).

The inability to influence the decisions of machines and to choose specific datapoints to carry out the analyses necessary to evaluate the significance of an action risks delegating responsibility to the machines themselves, "which will not decide the way would, but probably not even how we could accept that they do" (Wiener 1966, 228). The concept of individual responsibility makes it possible to address the issue of risk factors for individuals in the use of big data, further expanding the concept of technical responsibility towards the environment (Jonas 1979)—see also Sect. 1.2.3. If man, way back in Sophocles' Antigone chorus, was able, through his work, to learn, changing and overcome the limits of nature, with current technologies, this power of "torturing nature to discover its secrets" (Bacon 1620) has drastically expanded. For this reason, this responsibility must also be carefully evaluated in the field of ICTs, considering their potential impact—see also Chiesa (2017a, 2013a), Pagani et al. (2015)—so as not to delegate such a delicate passage to the intelligence of machines.

Referring to the work of Mayer-Schönberger and Cukier in 2013, there are four key points on which big data risk making people vulnerable to the misuse of technologies:

- the protection of privacy, in a context in which it is not yet sufficiently confirmed that big data favour de-anonymisation, so much so that an increasing amount of data is collected and combined. This involves a shift from group identities and user typing to the identities of each individual;
- the need to avoid penalising people on the basis of their propensities. There is a "possibility to use forecasts that emerge from big data on people to judge and punish them even before they act. And it is the denial of the concepts of equity, justice and free will" (Mayer-Schönberger and Cukier 2013, 204). This concept is the evolution of Wiener's thoughts on delegating responsibility to machines. In the world of big data, this problem is augmented by the possibility of depriving individuals of freedom of choice, acting on the collectivisation of human choices by erasing the related concept of free will. It will be necessary to avoid the development of paternalistic states or organisations capable of punishing people in advance and firing them if they suffer from probable genetic diseases or risks of criminal behaviour. This is linked to philosophical and theological reflections on the responsibility of man and freedom of choice and could remind us of the thoughts of Cesare Lombroso;
- the risk of a data dictatorship "on the basis of which we elevate information to fetish status" (Mayer-Schönberger and Cukier 2013, 204). Big data is a tool, not

a definitive and absolute answer. For this reason, it is necessary to strengthen the awareness and knowledge of what we want so that we do not forget the actual meaning of the data and information themselves;

- ownership and control of data, which can be used to manipulate choices and repercussions or can facilitate the use of human beings as mechanisms and which, "f mismanaged, risk becoming a tool at the service of the powerful" (Mayer-Schönberger and Cukier 2013, 205). Think of the *curse of McNamara*.

These premises allow the development of some schemes to identify decision-making methods on an urban scale, towards Smart City, in the perspective in which it is becoming a determining factor in the potential for action on societies and on virtual or real spaces.

6.2.1 Data, Cities, Properties, and Decisions

The analysis of the current strategies of the Smart City candidate cities highlights the need to create platforms of stakeholders, including administrations and local authorities, universities and research institutions, companies and industries, banks and financiers, and of available technologies, i.e. hardware, software and orgware. The risk of large industrial groups and major economic actors dominating urban policy decisions is a possible point of concern that should be addressed with particular attention when defining European and national funding strategies. At the same time, ICTs make it possible to increase the potential participation of the population, in a process that risks, however, fuelling a hyper-democracy based on a techno-political plebiscite (Occelli and Staricco 2002). We can say that the two phenomena are likely to take place alongside one another: an increase in the decision-making power of large groups can lead to an increase in hyper-democratic solutions capable of shifting attention from policy-related decisions to the choice of ornamentation. The technological repercussions, derived from a reduction in the role of public authorities and cultural decision-makers, could be numerous. On the one hand, the research and development phase would move away from the definition phase of the Smart City, in which subjects propose technologies and ready-developed "turnkey" packages. On the other hand, they would involve citizens and local businesses only later, i.e. in the implementation phase of specific actions. In addition, universities, research centres, cultural spaces, innovative small businesses and spin-offs would find it difficult to be part of this process of building the smart city, and at the same time, data and technologies would risk being restricted to big companies or possibly small businesses capable of working their way into the decision-making folds. This last concept is particularly relevant when interfacing with proprietary protocols, such as the field of home automation for example. Specifically, the technologies favoured in the ICT sector and in home automation would be those proposed by the strongest industries in the sector, tending to create proprietary languages based on minimally interoperable platforms. Yet, alongside top-down systems, low-cost, easy-to-manage open source and open hardware tools which, if applied according to standardised protocols, allow the creation and management of widespread, connected and shared digital systems that could easily find application in the creation of open platforms, exist and others are being developed. The way platforms are managed can also lead to different approaches between ICTs and the development of the Smart City. Particularly in the case of top-down solutions, there is a risk of incurring linear approaches that have a direct cause and effect between the application of ICTs and their impact on the city (technological determinism) or of falling into a futuristic utopianism that sees smart technologies as the solution to every problem (Occelli and Staricco 2002, 20). We need to work towards the adoption of more complex models, governed by feedback and economic-social networks, which link ICTs, the impact on the urban fabric and interconnections with the social and economic fabric.

The development of it has at least two main effects linked to the two terms that make up the abbreviation: technologies, on which we are focusing, and information, on which, we are going to be focusing in the years to come (Mayer-Schönberger and Cukier 2013). Each of the two impacts has and reinforces specific implications on policies, platforms and design and decision-making methods.

The main interconnections between information, technological management platforms and visions for the future of our cities are described in the diagram in Fig. 6.12, which mainly analyses technology-related aspects.

The diagram allows the identification of four macro-categories which, starting from different organisations of Smart City urban platform, involve different ways of processing and acquiring data, creating models and applying technologies. In particular, the abscissa axis measures the involvement and weight of the administration and the collective or individual interest in the application of current information

Fig. 6.12 Urban visions, data and accessibility, a four axes scheme. Re-elaborated from Chiesa (2013b)

technologies, while the ordinate axis analyses the degree of closure and openness of the technologies proposed by the Smart City urban platform. A top-down approach involves the predominance of technologies consolidated and developed by a small number of partner industries, often multinationals, and leaves little room for specific need-driven flexibility on the one hand, and for updating and expanding systems on the other. As far as data is concerned, the top-down approach envisages centralised data management, organised by an administrative body or a collective body, in compliance with an approach that refers to the concept of Orwell's Big Brother, or managed by private companies, according to an individualist drift that is often articulated around the concepts of technical anti-comprehension and flexible man (Sennet 1999, 2006).

The bottom-up approach, on the other hand, sees the application of technologies, partly consolidated, partly open source and partly open hardware, with the direct involvement of users or artisan knowledge [in the Sennettian sense, Sennet (2008)] of a local nature, which can be conformed to suggested packages, definable as communitarian, or to private needs that have no ramifications, i.e. individual, involving universities and research and development companies.

The bottom-up approach involves individual management of data or a transfer of output data to administrative subjects capable of making possible suggestions for input interventions and strategic planning actions.

The right side of the diagram can be classified as anarchist, the left side as organised. In the first case, a widespread development of the city will be favoured [see the images in Basilico (2005)], while in the second case, a widespread and reticular development (Detragiache 2003) will involve the passage from settlement dispersion to an "organised reticulation, which controls the consumption of space and safeguards the quality of the environment" (Occelli and Staricco 2002, 9). The upper part of the diagram can be called oligarchic, the lower part democratic. The combination allows the definition of the top down-community approach as technocratic dirigisme, the top down-individual approach as survival of the fittest, the open source-individual as a refuge/shelter, in the double meaning of environmental refuge or space of the Pioneer. The Pioneer is understood as a national hero "who worked more than everyone else to convert into ready cash [the natural heritage, which] we have exalted as if he were the creator of those riches which, in actual fact, he plundered and squandered" (Wiener 1966) exploiting what is known as fifth freedom. The last, open-source community, box is defined as the field of strategic planning-community in which are evident some ordering principles and some "lines of force of the urban form [...] consistent with the morphological peculiarities of the territory" (Occelli and Staricco 2002, 9).

It is therefore necessary to assess, during the creation and implementation of platforms and strategic plans for the transition to Smart City, the implications that ICT and digital technologies, in general, have and will have on the social and urban fabric and on city planning in the dual role of enabling and co-evolving these technologies.

Concentrating on the Information aspects of IT, the focus shifts mainly to the implications that the big data revolution will have on the social, urban and architectural fabric. From the decision-making point of view, it is important to analyse the main types of players and companies specialising in different areas of the big data value chain, which "can be differentiated according to the value offered: data, skills and ideas" (Mayer-Schönberger and Cukier 2013, 169). In particular, the data domain is characterised by players who own, produce or have the ability to access data, such as Twitter, Facebook, Google and Public Administrations. The area of expertise is populated by consulting firms, *data specialists*, who gain valuable information from the data, which they give to the owners of the data. This specific area will tend to change radically over time with the redefinition of the role of expert, which is likely to happen in the coming years. Finally, the sphere of ideas is made up of players capable of developing creative models for the use of data, which tend to analyse big data and also provide cultural orientation. Figure 6.13 outlines the urban visions deriving from open or proprietary data approaches with respect to the three main categories of player involved in big data. The larger pyramid represents an open public approach, in which the management of processes and ideas is mediated by decision platforms coordinated by public administrations or by bodies that act in the collective interest. These analyses take place in public scientific research centres such as universities and the data are open source. This scenario fuels the enabling power of ICTs in a population of citizens by introducing the concept of building citizenship. The medium-sized pyramid shows an intermediate view, where ICTs have enabling and limiting roles depending on location and decision-making fields, pointing to specific choices. In this scenario, decision-making platforms and actors see an alternation of choices aimed at favouring the individual or the community,

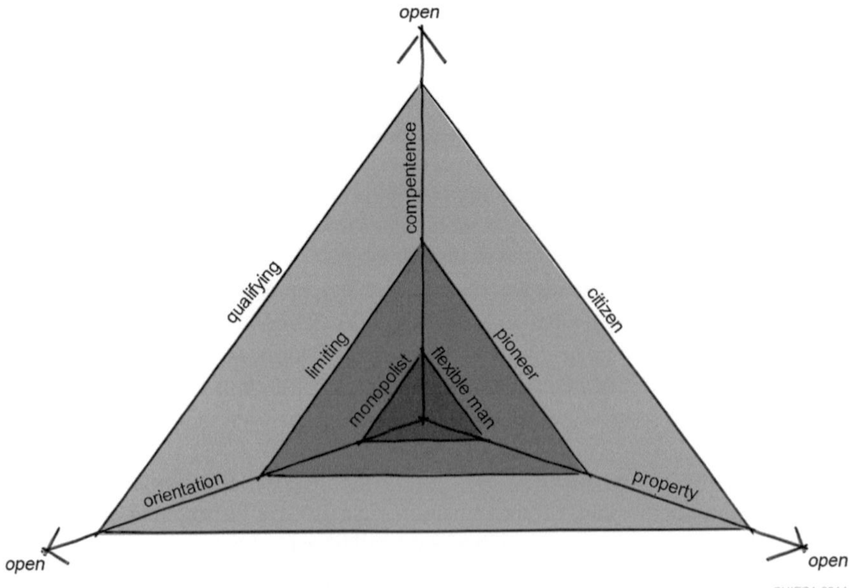

Fig. 6.13 Urban visions, data and stakeholders (schematic representation based on the information side of IT)

generating a space for action for the pioneer who is able to assert himself. The smaller pyramid shows a completely monopolistic or strongly oligarchic view of the big data decision-makers. This area could represent a space in which individuals behave like ants, without being questioned in terms of ownership, decision making and analysis. People strengthen their flexibility and liquidity in cities where specific knowledge and individuality are secondary.

Figures 6.14, 6.15 and 6.16 show intermediate diagrams, where an unrestricted axis relates to visions in which at least one of the players is exclusive or confidential. The prism with free competence and orientation and restricted ownership, represents the case in which the data belong, for example, to a multinational company that may or may not leave the possibility to use them to decision-makers and public analysts or third parties. It can be a scenario in which public administrations use Google's own data to optimise urban transport. The case of open ownership and orientation and individual competence, can build the example of a public administration that commissions a private company to analyse its data according to specific requirements. This is the transposition, within the scope of data, of the use of private agencies to manage fines. There are, of course, risks arising from a lack of control of residual or parasite uses of data or their dissemination by the agency responsible for them. The case of public ownership and competence, and private orientation represents an individualistic forcing of the collective protection mechanisms.

The progressive increase in the weight of information on companies, cities and architectures involves considerable risks, as already mentioned in this chapter. It is essential to bear in mind that data can lead to incorrect readings, due to mislead-

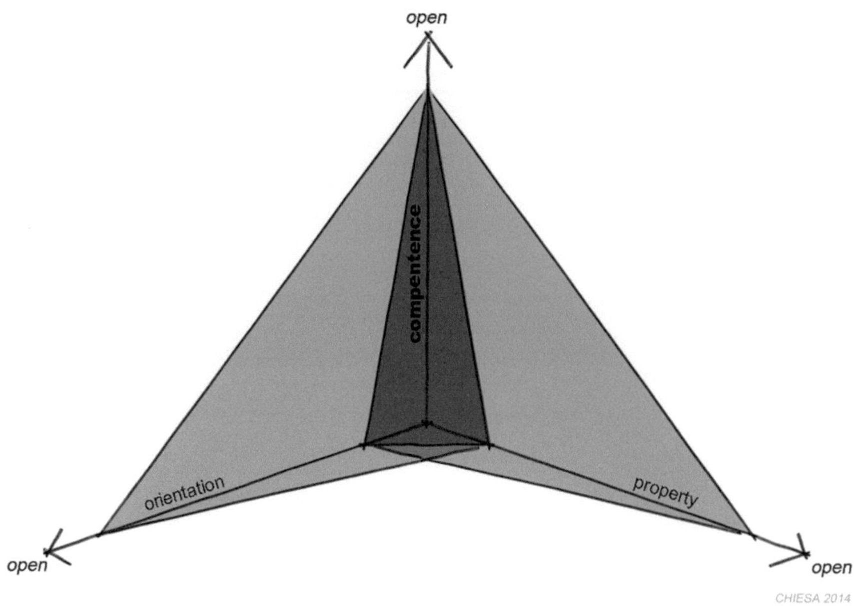

Fig. 6.14 Urban visions, data and stakeholders (open axis concerning competence)

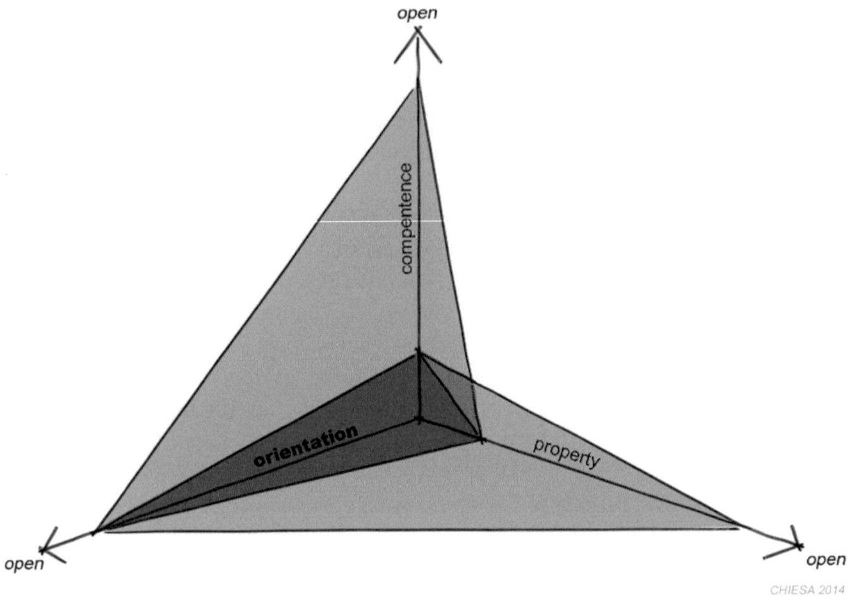

Fig. 6.15 Urban visions, data and stakeholders (open axis concerning orientation)

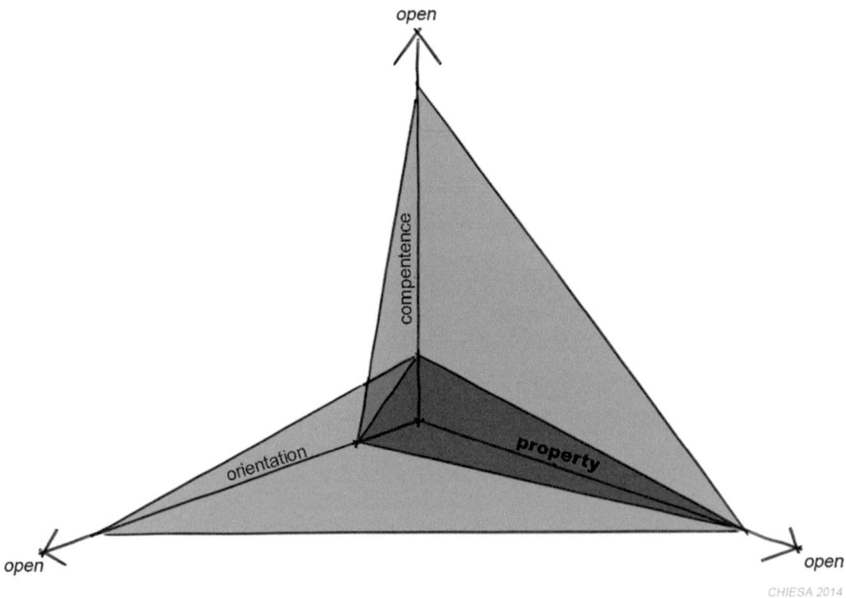

Fig. 6.16 Urban visions, data and stakeholders (open axis concerning data property)

ing or manipulated use, incorrect or improper methods and choices of analysis, the use of inconsistent, poor quality, non-ontological datasets, based on an incorrect interpretation of metadata, an incorrect choice of parameters and indicators or an excessive focus on calculations, a supporting and non-objective tool, which could increase our vulnerability to black swans (Taleb 2009, 264). In other words, it is necessary to avoid the mistake of Icarus "who, blinded by his technical skill at flying, used it carelessly and fell into the sea" (Mayer-Schönberger and Cukier 2013, 229). It is also essential to identify, from the point of view of an urban and strategic decision-making platform, the boundary and the degree of uncertainty that links the forecasts made by big data with the need to ensure freedom of choice and individual responsibility. In this sense, the fourth Cartesian quadrant of the scheme of Fig. 6.12, also identifies the State and the public administration that act in a paternalistic and preventive way. In order to prevent the centrality of data and information influencing the platforms of design and decision-making like "black boxes", of which we know the forecast but not the metadata related to it, including reliability, calculation methods and processors, it is essential to have technological and theoretical tools capable of managing and influencing the methods of modelling, materialisation, choice and design in the digital age. Data is a tool, not a decision maker. It is up to designers and those who possess technical knowledge to acquire the skills necessary to become algorithm specialists and make sure that they play an infinite and not a finite game (Carse 1986).

References

Augé M (2004) Rovine e macerie. Il senso del tempo. Bollati Boringhieri, Torino [or (ed) (2003) Le temps en ruines. Editions Galilée, Paris]

Bacon F (1996, [1620]) Nuovo organo. RCS, Milano

Basilico G (2005) Scattered city. Baldini Castoldi Dalai editore, Milano

Carse JP (1986) Finite and infinite games. Ballantine Books, New York and Toronto

Chemollo A, Orsenigo F (2004) Senza Posa. Marsilio, Venezia

Chiesa G (2013a) La città digitale, dai sensori ai modelli: Piattaforme interconnesse per la città del futuro. In: Di Giulio R et al (eds) Strategie di riqualificazione urbana: Rigenerazione e valorizzazione dell'edilizia sociale ad alta densità abitativa del secondo Novecento. Quodlibet, Macerata, pp 110–117

Chiesa G (2013b) M.E.T.R.O. (Monitoring energy and technological real time data for optimization) innovative responsive conception for city futures. Ph.D. thesis, Politecnico di Torino, Torino

Chiesa G (2017a) Explicit-digital design practice and possible areas of implication. Techne 13:236–242

Chiesa G (2017b) Social indicators to localize renewable energy sources considering their visual impacts. Energy Procedia 122:529–534

Chiesa G, Di Gioia A (2011) Rappresentare il territorio della contemporaneità: la fotografia ambientale come supporto all'analisi territoriale. Planum, 11 pp

Chiesa G, La Riccia L (2011) Fotografia e indicatori del paesaggio. Lo spazio dell'abitare. Planum, 10 pp

Chiesa G, La Riccia L (2013) Dalla rappresentazione alle rappresentazioni di paesaggi e territori. Planum 27(2)

Chiesa G, La Riccia L (2016) Tools and methods for evaluating and designing the perceived landscape. 3D-GIS, viewshed analysis, big data. In: Pagani R, Chiesa G (eds) Urban data. Tools and methods towards the algorithmic city. FrancoAngeli, Milano, pp 159–198
Detragiache A (ed) (2003) Dalla città diffusa alla città diramata. Franco Angeli, Milano
Floridi L (2013) The onlife manifesto. Seminary, Centro Nexta about Internet & Società. Politecnico di Torino, DAUIN, Turin, Italy, 24 May 2013
Ghirri L (1997) Lo sguardo inquieto. Un'antologia dei sentimenti. In: Costantini P, Chiaramonte G (eds) Niente di antico sotto il sole. SEI, Torino
Girardin F et al (2007) Understanding of tourist dynamics from explicitly disclosed location information. In: 4th international symposium on LBS and telecartography, Hong-Kong, China
Jonas H (1979) Das Prinzip Verantwortung: Versuch einer Ethik für die technologische Zivilisation. Frankfurt [ITA trans: (1990) Il principio di responsabilità. Einaudi, Torino]
Kuhn TS (1969) La struttura delle rivoluzioni scientifiche. Einaudi, Torino [or (ed) (1962) The structure of scientific revolutions. University of Chicago Press, Chicago]
La Riccia L (2017) Landscape planning at the local level. Springer, Berlin
Lupano M (1989) Fotografare I luoghi fotografare le architetture. Intervista di Mario Lupano a Luigi Ghirri. In: Ghirri L (ed) Paesaggio italiano, Electa, Milano, pp 10–12
Mayer-Schönberger V, Cukier K (2013) Big data. Una rivoluzione che trasformerà il nostro modo di vivere e già minaccia la nostra libertà. Garzanti, Milano [or (ed) (2013) Big data: a revolution that will transform how we live, work, and think. Houghton Mifflin Harcourt, Boston]
Mitchell WJ (2005) Construction complexity. In: Martens B, Brown A (eds) Computer aided architectural design futures 2005. Springer, Netherlands, pp 41–50
Nielsen M (2012) Le nuove vie della scoperta scientifica. Come l'intelligenza collettiva sta cambiando la scienza. Einaudi, Torino [or (ed) (2012) Reinventing discovery: the new era of networked science. Princeton University Press, Princeton]
Occelli S, Staricco L (2002) Nuove tecnologie di informazione e di comunicazione e la città. FrancoAngeli, Milano
Oxman R (2006) Theory and design in the first digital age. Des Stud 27:229–265
Pagani R, Chiesa G, Tulliani J-M (2015) Biomimetica e Architettura. Come la natura domina la tecnologia. FrancoAngeli, Milano
Sennet R (2008) The craftsman. Yale University Press, New Haven
Sennett R (1999) L'uomo flessibile. Feltrinelli, Milano [or (ed) (1998) The corrosion of character: the personal consequences of work in the new capitalism. Norton, New York, London]
Sennett R (2006) Il declino dell'uomo pubblico. Bruno Mondadori, Milano [or (ed) (1976) The fall of public man. Knopf, New York]
Sevtsuk A, Amindarbari R (2012) Measuring growth and change in East-Asian cities. Progress report on urban form and land use measures. The World Bank, City Form Lab, Singapore
Taleb NN (2009) Il Cigno nero, come l'improbabile governa la nostra vita. Il Saggiatore, Milano [or (ed) (2007) The black swan: the impact of the highly improbable. Random House and Penguin Book, New York]
Tosoni P (1992) Il gioco paziente. Biagio Garzena e la teoria dei modelli per la progettazione. Celid, Torino
Weinberger D (2012) La stanza intelligente. La conoscenza come proprietà della rete. Codice edizioni, Torino [or (ed) (2011) Too big to know: rethinking knowledge now that the facts aren't the facts, experts are everywhere, and the smartness person in the room is the room. Basic Book, New York]
Wiener N (1966) Introduzione alla cibernetica. L'uso umano degli esseri umani. Editore Boringhieri, Torino [or (ed) (1950) The human use of human beings: cybernetics and society. Houghton Mifflin Company, Boston]
http://cityform.mit.edu/, last view Apr 2019

Conclusions

The book introduces a small part of the vast and constantly growing horizon of ICT and IT pervasiveness in architecture and urban planning. However, it proposes an interpretative key which focuses on three areas (eras) that, though interconnected and generally interrelated in reality, can define three keys of innovation that can be connected to many new macro-knowledge and digerate professional figures. On the one hand, the relationship with advanced modelling, virtualization and real-time rendering horizons is analysed while, on the other hand, the relationship with mass-customized materialization, 3-D printing, and contour manufacturing that leads to Industry 4.0 is analysed. Finally, the field of data analysis, data production and IoT (Internet of Things), which links all areas, but also opens up new and unexpected horizons and professional fields in building design and operation is introduced. The boundaries between ICT and architecture are narrowed. Several of my ICT students have, in fact, demonstrated an incredible ability to solve design problems thus paving the way for an entry into the world of architecture, while some of my architecture students have dealt with ICT problems in their dissertations thus working in a totally hybridised reality. As has been stated by several authors this is the time to surf the ICT innovation wave in order to avoid the risk of the extinction of architects. New tools are opening up incredible new horizons, but have to be used in a conscious way. The design culture has in fact to open out to ICT culture and knowledge because methodologies are not always the same as in the paper-based or traditional CAD worlds. Never before has it been more vital to understand and manage technical potentialities in both the architectural and the ICT fields. ITs allow us to improve the architectural world, but like any technology they can become a nightmare or lead to wrong or inaccurate results if not used well. As scientists, designers and technicians, we are part of this innovation and we are linked together through joint responsibility.

As can been envisaged by the proposed samples, the real limit to the integration of IT and ICT within the architectural sector is mainly function of our imagination, vision, and ability to manage interdisciplinary knowledge. This innovation is showing a very large potential for increasing the efficiency in the entire building

G. Chiesa, *Technological Paradigms and Digital Eras*,
PoliTO Springer Series, https://doi.org/10.1007/978-3-030-26199-3

process, from early-design stages, to maintenance and building operational phases, including construction processes, and end of life. All of these aspects can be coupled with specific digital implications. Connected network of sensors may feed neural networks to predict and operated in advanced building systems e.g. for comfort purposes; computational tools may be used to optimise design choices, produce mass customized building forms and components using materials in a totally innovative way. Nevertheless, it is essential to also consider the social implications of ICT pervasiveness in our society and in living spaces, suggesting conscious visions and having the ability in being designers, architects, engineers in the digital eras by also considering our responsibilities.

MIX
Papier aus verantwortungsvollen Quellen
Paper from responsible sources
FSC® C105338

If you have any concerns about our products,
you can contact us on
ProductSafety@springernature.com

In case Publisher is established outside the EU,
the EU authorized representative is:
Springer Nature Customer Service Center GmbH
Europaplatz 3, 69115 Heidelberg, Germany

Printed by Libri Plureos GmbH
in Hamburg, Germany